Nonequilibrium Ecology

Ecology has long been shaped by ideas that stress the sharing of resources and the competition for those resources, and by the assumption that populations and communities typically exist under equilibrium conditions in habitats saturated with both individuals and species. However, much evidence contradicts these assumptions and it is likely that nonequilibrium is much more widespread than might be expected. This book is unique in focusing on nonequilibrium aspects of ecology, providing evidence for nonequilibrium and equilibrium in populations (and metapopulations), in extant communities and in ecological systems over evolutionary time, including nonequilibrium due to recent and present mass extinctions. The assumption that competition is of overriding importance is central to equilibrium ecology, and much space is devoted to its discussion. As communities of some taxa appear to be shaped more by competition than others, an attempt is made to find an explanation for these differences.

KLAUS ROHDE is Professor Emeritus at the University of New England, Australia.

Nonequilibrium Ecology

KLAUS ROHDE

School of Environmental Sciences and Natural Resources Management, University of New England, Australia

Shaftesbury Road, Cambridge CB2 8EA, United Kingdom

One Liberty Plaza, 20th Floor, New York, NY 10006, USA

477 Williamstown Road, Port Melbourne, VIC 3207, Australia

314–321, 3rd Floor, Plot 3, Splendor Forum, Jasola District Centre, New Delhi – 110025, India

103 Penang Road, #05–06/07, Visioncrest Commercial, Singapore 238467

Cambridge University Press is part of Cambridge University Press & Assessment, a department of the University of Cambridge.

We share the University's mission to contribute to society through the pursuit of education, learning and research at the highest international levels of excellence.

www.cambridge.org
Information on this title: www.cambridge.org/9780521674553

© Klaus Rohde 2005

This publication is in copyright. Subject to statutory exception and to the provisions of relevant collective licensing agreements, no reproduction of any part may take place without the written permission of Cambridge University Press & Assessment.

First published 2005

A catalogue record for this publication is available from the British Library

ISBN 978-0-521-85434-4 Hardback
ISBN 978-0-521-67455-3 Paperback

Cambridge University Press & Assessment has no responsibility for the persistence or accuracy of URLs for external or third-party internet websites referred to in this publication and does not guarantee that any content on such websites is, or will remain, accurate or appropriate.

Für Ursula

Contents

Acknowledgements

I thank Peter Rohde who wrote the program for generating Figure 3.3 and helped with scanning the other figures. He also drew my attention to the book by Wolfram (2002) and made helpful comments on the sections dealing with cellular automata. I am grateful to anonymous referees for valuable comments on the entire manuscript. I wish to thank the following colleagues for commenting on sections of the manuscript: Josef Alvermann, Stuart Barker, Stuart Cairns, Peter Clarke, Howard Cornell, Hugh Ford, Nick Gotelli, Illka Hanski, Chris Nadolny, Ross Robertson, Peter Sale, Diane Srivastava, Gimme Walter. David Jablonski gave advice on fossil species diversity, Stuart Cairns advised me on literature dealing with kangaroo ecology and population biology, Bernard Seret and Tim Berra on literature dealing with fish diversity in Madagascar, Peter Sale and Ross Robertson on reef fish literature, Peter Clarke and Chris Nadolny on tropical rainforest literature, and Hugh Ford on literature dealing with recent extinctions.

The following kindly permitted the use of figures: Andrew Allen, Stuart Cairns, Howard Cornell, Don DeAngelis, James Gillooly, Illka Hanski, M. A. Hixon, David Jablonski, Kazuo Kawano, Armand Kuris, Sergio Navarette, Robert Poulin, Mark Ritchie, Michael Rosenzweig, Peter Sale, Jianguo Wu, and the editors of Oikos and Ecography.

I also wish to thank John Hooper of the Queensland Museum, Brisbane, and Charlie Veron of the Australian Institute of Marine Science, Townsville, Queensland, for the cover photos.

Introduction

Ecology has long been shaped by ideas that stress the sharing of resources and the competition for those resources, and by the assumption that populations and communities typically exist under equilibrium conditions in habitats saturated with both individuals and species. This view can be traced back to Linnaeus, who considered an equilibrium in nature; Adam Smith, who contributed the idea that competition can lead to equilibrium in a community; and Malthus, who suggested that greater growth in demand than in supply would lead to competition for limited resources. Among well known ecologists, Hutchinson (1948) took it for granted that stability (owing to "self-correcting mechanisms") is characteristic of most ecological systems and permits their persistence, and, according to Dobshansky (1957, cited by Cooper 2001): "natural selection, and hence the evolutionary process, are the outcome of competition; and therefore are governed by density-dependent factors." Some ecologists were always aware of the possibility of nonequilibria, but the majority ignored it, especially in connection with theory in ecology. In several widely used older ecological texts, competition and equilibria are discussed in depth, but nonequilibria are not mentioned at all or only in a very cursory fashion (e.g., Pielou 1969; MacArthur 1972; Cody and Diamond 1975; Ehrlich *et al.* 1977). This has changed somewhat in recent years, particularly in population ecology (e.g., Chesson and Case 1986; Diamond and Case 1986; DeAngelis and Waterhouse 1987; Krebs 2001). Nevertheless, many workers still seem to be pre-occupied with looking for evidence of competition and equilibria. In the 20 review articles in Fenchel (1999), which deal with aspects of ecology supposed to be of major current and future interest, only one contains the term nonequilibrium as important for future research in the list provided by the editors (Lindström *et al.* 1999). Negative results, i.e., results failing to provide evidence for competition, seem to be reported as rarely as they were 20 years ago, when Pickett (1980) and Price (1980, 1983, 1984) drew attention to this lack of information. Rosenzweig and Ziv (1999) state "Theory suggests that higher diversity should shrink niches, allowing the coexistence of more species locally," which clearly is an assumption of equilibrium ecology, and even some of the basic assumptions of Hubbell's (2001) neutral theory of biodiversity are not different from those of equilibrium ecology. He assumes, for example, that communities are saturated with individuals leading to a zero-sum game, that numbers of individuals of one species can increase only at the expense of individuals of other species, and that there is saturation with species.

There is no recent book that focuses on nonequilibrium aspects of ecology. This book aims at filling this gap. After a brief outline of concepts and problems, an outline of historical milestones in the evolution of ideas, and the description of some empirical studies demonstrating equilibrium and nonequilibrium, I define the major problem discussed in this book (the relative importance of equilibrium and nonequilibrium), using the arguments in favour of prevailing equilibrium conditions in nature provided by Rosenzweig (1995). I then discuss evidence, including some detailed examples, for nonequilibrium and equilibrium in populations (and metapopulations), in extant communities and in ecological systems over evolutionary time, including nonequilibrium due to recent and present mass extinctions. The assumption that competition is of overriding importance is central to equilibrium ecology, and much space is devoted to its discussion. An autecological comparison of two parasite species, using evidence from phylogeny, morphology, life-cycle, and ecological studies, attempts to evaluate whether equilibrium assumptions, in particular assumptions on interspecific competition, can explain adaptations and extant patterns of distribution. Finally, because communities of some taxa appear to be shaped more by competition than others, an attempt is made to find an explanation for these differences.

The bias towards equilibrium ecology is at least partly due to a bias in selecting models for study. Indeed, there has been a strong bias towards pest and some other insect species occurring at high densities, and towards birds and terrestrial vertebrates, which in many cases live in communities that are to a large degree saturated and exposed to competition for limiting resources. This book discusses some such examples, but, in addition, uses many examples from systems that have often been neglected, i.e., parasitic and aquatic systems. Species in such systems represent the majority of eukaryotic species on Earth and their study may give a more balanced view of how ecological systems work. Also, many parasites live in well defined habitats and communities with an almost unlimited number of replicas, which can easily be manipulated in natural experiments, i.e., they are excellent objects for ecological studies.

1 · *Concepts and problems*

Concepts of equilibrium (balance of nature) and nonequilibrium

The concepts of equilibrium/nonequilibrium have been used differently by different authors, as will be shown in the following selected examples.

Pianka (1974), in the second edition of his widely used *Evolutionary Ecology*, makes the case succinctly for equilibrium in ecological systems (modified somewhat in later editions). The main points listed by him are that:

(1) ecological systems and their components have been shown to be in dynamic equilibrium near steady states in many studies;
(2) in communities, production and respiration "must ultimately balance";
(3) even nonclimax communities, which have not reached a steady state, are probably "in some kind of equilibrium," determined by the frequency of disturbances and destruction of other successional stages and the rate of successional change;
(4) in most communities, rates of energy influx and outflow in each trophic level balance out exactly;
(5) on islands, immigration and extinction of species are balanced;
(6) in populations, over long periods, birth rates equal death rates; and
(7) prey–predator and similar pairs must be "in some sort of ecological and evolutionary balance to coexist with one another over any period of time."

The assumption that competition plays a central role in ecology is implicit in Pianka's discussion. This notion has since been explicated by Chesson and Case (1986), who define the assumptions of "classical competition theory" as follows:

(1) life history characteristics of species are adequately summarized by the per capita growth rate of species;

(2) deterministic equations are sufficient to model population growth, and environmental fluctuations need not be considered;
(3) the environment is spatially homogeneous and migration is unimportant;
(4) competition is the only important biological interaction; and
(5) coexistence requires a stable equilibrium point.

They further summarize the consequences of these assumptions, i.e., that *n* species can coexist only if there are no fewer than *n* limiting resources, and that there is a limiting similarity between species: species evolved in response to interspecific competition. Assumption 4 (above) is relaxed when predators are present, i.e., *n* species may coexist when fewer than *n* resources are present. If the environment favours different species in different patches, *n* species may coexist in at least *n* patches, even if the species use the same limiting resource.

The authors then contrast these assumptions with those of nonequilibrium situations. Nonequilibrium is seen as "any situation where species densities do not remain constant over time in each spatial location." Even if fluctuations occur at small spatial scales that level each other out at larger scales, an explanation would still be a nonequilibrium explanation if the fluctuations are an essential part of the explanation. Chesson and Case (1986) discuss the following four points that deviate from classical equilibrium assumptions:

(1) populations are not at a point equilibrium but competition still occurs continuously and is important; this may permit more than *n* species to coexist on *n* resources (e.g., Armstrong and McGee 1980);
(2) fluctuations in population density or environmental variables are dominant, population dynamics may be density-independent;
(3) means and variances of environmental fluctuations are not constant over time; and
(4) populations are random-walking, but time to extinction is so long that species persist over a long time ("slow competitive displacement").

Concerning nonequilibrium, Cappucino (1995) argues that only randomly walking populations are unambiguously nonequilibrial. All other usages are misplaced. Thus, the most commonly used meaning of nonequilibrium in populations refers to situations where local populations do not trend towards a point equilibrium (density-vagueness of Strong [1984]; stochastic boundedness of Chesson [1978]). However, according to Cappucino, fluctuations in such populations are bounded and therefore regulated and not nonequilibrial. Moreover, if all subpopulations of a metapopulation are randomly walking, the metapopulation as a whole must also randomly walk to extinction (Chesson 1981).

An important discussion of equilibrium in nature is provided by Cooper (2001), who looks critically at the various arguments given by different authors in favour of equilibrium. He distinguishes two "balance of nature" arguments in population/community ecology. The first argument is based on the assumption that there is a strong tendency towards constancy in population size. But how much variability is allowed before the population is no longer considered to conform to this alleged constancy? The second argument has been developed to overcome this difficulty. It simply assumes that the balance is represented by populations tending to persist. But this persistence must be the result of some kind of regulation. The reason is that random-walking populations must become extinct over time; therefore, those that do persist must be regulated in some way, in the sense that they must display statistical stability over time.

Cooper refers to several authors who have proposed explanations of stability (i.e., May 1973; Chesson 1981, 1982; Murdoch 1994; Dennis and Taper 1994; Turchin 1995). The last two papers discuss the problem in terms of stationary probability distributions (SPD) in population density–time series. Population densities are assumed to fluctuate around a mean density level, and the variance of fluctuations is bounded in the long term. Turchin (1995) defines equilibrium (or being regulated) as existence of an SPD. Since most populations persist, i.e. have an SDP, they must be in equilibrium. In other words, they must be regulated. Also, according to this argument, regulation implies density dependence, because populations would either become extinct or increase *ad infinitum*, if density-dependent factors did not operate (May 1986). Most likely, density-dependent factors are competitive ones, at least if predation and other such influences do not limit population density to a level below which competitive effects can become operative. However, as pointed out by Cooper (2001) and others to whom he refers, all populations become extinct sooner or later, and one would have to demonstrate that populations persist longer than expected if they simply executed random walks. Furthermore, a tendency to return may only be a necessary but not a sufficient condition for equilibrium (or regulation), i.e., empirical evidence for the occurrence of regulation has still to be given. In other words, Cooper argues that the inclination of many (if not most authors) to use a-priori arguments for a balance in nature (equilibrium, regulation) is ill advised.

History of equilibrium and nonequilibrium ecology: some milestones in the evolution of ideas

This account concentrates on the development of some important ideas in the equilibrium-versus-nonequilibrium debate, it does not attempt to give a complete history of equilibrium and nonequilibrium ecology. An early account of species equilibrium was given by Wilson (1969), and an excellent history of population ecology is given by Kingsland (1995). Harris (1986) provides some information on the development of equilibrium and nonequilibrium theory in ecology, and Egerton (1973) has given a detailed discussion of the history of the concept of balance in nature (see also Pimm 1991). According to Walter and Patterson (1995), "an uncritical and scientifically unsupported belief in the strength and central role of competition (both interspecific and intraspecific) in evolution and ecology is traceable to Darwin (1859) and beyond." Darwin wrote in his *Origin of Species* (1859, cited by Silverton 1980) that "Battle within battle must be continually recurring (in nature) with varying success; and yet in the long-run the forces are so nicely balanced, that the face of nature remains uniform for long periods of time, though assuredly the merest trifle would often give the victory to one organic being over another." The "beyond" Darwin includes Linnaeus, who considered an equilibrium in nature; Adam Smith, who contributed the idea that competition can lead to equilibrium in a community; and Malthus who suggested that greater growth in demand than in supply would lead to competition for limited resources (see Hengeveld and Walter 1999 for a further historical discussion). Hutchinson (1948) takes it for granted that stability (owing to "self-correcting mechanisms") is characteristic of most ecological systems and permits their persistence. Or, as stated by Dobshansky (1957, cited by Cooper 2001) "natural selection, and hence the evolutionary process, are the outcome of competition; and therefore are governed by density-dependent factors." This assumption is the basis for Orians' (1962) opposition to the views of Andrewartha and Birch (1954), who did not recognize the predominant role of competition and the general existence of equilibria, and who were therefore supposed to have left ecology without a basis in the theory of natural selection.

As mentioned previously, Hutchinson (1948, 1959) believed that competition is a major factor determining species diversity and patterns in ecological communities, but he nevertheless recognized in 1961 that there are many more phytoplankton species in lakes than allowed by the classical competition theory (the "paradox of the plankton"). He suggested that nonequilibrium processes may provide an explanation for this, particularly those associated with temporal variations in lakes. Among eminent ecologists, Gleason (1926), Ramensky (1926), and much later, Whittaker (1967) were of the opinion that nonequilibria are important, and as early as 1952, von Bertalanffy concluded that living systems are open systems characterized by a continuous flow of substances and energy across their boundaries, that cannot establish true equilibria as found in closed systems.

Significant contributions to the establishment of what can be called nonequilibrium ecology are provided by Andrewartha and Birch (1954, 1984), and also Andrewartha (1970). Andrewartha and Birch, based on their extensive and intensive studies of many natural populations, conclude that too much emphasis on competition is fallacious. Most natural populations never become sufficiently dense to use a great proportion of the resources that they require. Density-dependent factors and competition therefore do not become operative (for details see p. 62).

In 1974, Levin and Paine published a seminal paper in which they showed that disturbance increased environmental heterogeneity, preventing local patches from ever achieving equilibrium, but, according to Berryman (1987), the first explicit statement of the viewpoint of nonequilibrial ecology was given by Caswell (1978), who said "equilibrium theories are restricted to behavior at or near an equilibrium point, while nonequilibrium theories explicitly consider the transient behavior of the system." This still implies the existence of equilibrium points, but systems are rarely at or close to these points. As earlier shown experimentally by Huffaker (1958) and theoretically by Maynard Smith (1974), spatial dimensions can prevent a system from "reaching a closed deterministic solution." In Berryman's view, "drawing distinctions between equilibrium and nonequilibrium theories of populations is distracting and misleading." Instead, there needs to be a theory developed that would explain population behavior both close to and far from equilibrium.

Extensive discussions of the nonequilibrium aspects of ecology are given in *Evolutionary Biology of Parasites* by Price (1980), and in *A New Ecology. Novel Approaches to Interactive Systems* edited by Price *et al.* (1984). In the Introduction to the latter volume, Price *et al.* (1984) point out that looking at various textbooks on ecology gives the clear impression that interspecific competition is of great importance, whereas the opposite view, i.e., that interspecific competition may be unimportant or only one of several equally important processes, has found little credence. An important reason for this unbalanced view, in the opinion of these authors, is the fact that negative data (for example those on the absence of competition) are not considered worthy of publication. Furthermore, "many scientists feel compelled to fit data into some existing body of theory, and do not feel equally compelled to falsify theory." Thus, very little of the supposed evidence for interspecific competition has been tested objectively (see Connell 1980, and pp. 64–65). Price (1984) therefore proposes that ecologists should test several hypotheses ("paradigms") to elucidate simultaneously processes determining community structure. These hypotheses include the null hypothesis that species respond individualistically to selection pressures, without competing for resources, simply living where conditions are favourable (Gleason 1926, Ramensky 1926); the resource heterogeneity hypothesis, that the number of species in a community is positively correlated with the number of resources, and their abundance with the abundance of the resources needed by the species; the island or patch size hypothesis, according to which area alone determines the number of species; the time hypothesis, according to which species numbers are determined by (ecological or evolutionary) time; and the enemy impact hypothesis, which says that enemies (e.g., parasites or predators) limit population sizes below the level at which resources become limiting. But not only should the relative importance of the various hypotheses be examined, but also the mechanisms involved. Hypotheses may be difficult to test for many communities and under many conditions. Price (1984, p. 374) lists characteristics of communities that make them suitable for hypothesis testing.

Price (1980) devotes a large chapter in his *Evolutionary Biology of Parasites* to "nonequilibria" in populations and communities, and other chapters also contain relevant information on this subject. In particular, the chapter on "Ecological niches, species packing, and community organization" gives a comprehensive discussion of nonequilibrium in parasite communities. Equilibrium is a population state in which birth and death rates are equal, i.e., where there is no population growth (MacArthur and Wilson 1967, May 1973), or – more realistically in stochastic environments – where population size fluctuates around an average with steady average variance (May 1973). If variance is high, systems are unstable, if it is low, they are stable. Price (1980) draws attention to the difference between parasites and predators or browsers: a parasitic individual typically spends much of its life in a single or perhaps two patches (host individuals), whereas the latter search many patches. Globally, for parasites, there may be equilibrium, i.e., the proportion of patches occupied may remain stable, but within a patch, nonequilibrium conditions are highly likely (e.g., orange mites studied by Huffaker 1958).

Furthermore, parasite distributions within host populations are usually overdispersed, leading to undercolonization of most host individuals. Price notes that disruptions in parasite–host systems commonly occur, resulting in nonequilibrium states. The complexity of biotic interactions characteristic of a parasitic way of life, according to Price, is the main factor generating nonequilibrium; (also May 1973: complexity decreases stability). Price discusses several examples in detail. In each of these examples, patchy resources (host individuals) lead to a low probability of colonization, and ephemeral patches result in a high probability of extinction. *Schistosoma* populations provide one of his examples. Here, patchiness of resources is indicated by the patchy distribution of water, the independent differentiation of snail host populations, highly specific parasite strains, and patchiness of snail and vertebrate host distribution. Ephemeral patches are due to temporary availability of water, high mortality, sterility or reduced fecundity of snail hosts, the dynamics of snail and vertebrate hosts, and developing host resistance.

However not only in populations, but also in communities of parasites, equilibrium conditions are unlikely, since the parasite populations of which they are comprised are all in nonequilibrium (Price 1980). Mechanisms leading to nonequilibrial communities are the same as for populations: if colonization probabilites are low and extinction probabilities high – which is usually the case – nonsaturation and nonequilibrium will result. Price discusses a number of examples and concludes that, if interspecific competition occurs in such systems, it is intermittent.

Strong (1984) introduced the term density-vagueness to describe parameters of birth and death rates that are only weakly explained by density. Sometimes, the density effect can be inferred only with imagination, or the correlation may be low (although statistically significant). Explanations other than density may be age, weather, migration, and others. He provides evidence that density dependence is not evident in a large number of studies and life-cycle stages, but he also points out why density dependence may sometimes be underestimated (but also overestimated). Nevertheless, for insects at least, density effects, where they occur, are often weak, intermittent, or discontinuous.

The belief that equilibrium conditions and competition are the major factors determining ecological processes, underlies and informs most ecological modelling. The concepts of evolutionary stable strategy (ESS), continuously stable strategy (CSS), and neighborhood invader strategy (NIS), for example, are based to a large degree on such an assumption (e.g., Apaloo 2003, further references therein).

Among major recent approaches, mentioned by Paine (2002), are the necessity to distinguish between open and closed systems (already recognized by von Bertalanffy 1952 and Levin and Paine 1974), and that "equilibrium may exist only in an abstract sense." Chesson and Case (1986) plead for a pluralistic approach: coexistence may be partly due to differences in resource use, partly because species respond differently to the environment, and partly because the advantage of one species over another is very small, leading to very slow competitive displacement; (see also the discussion in McIntosh 1987).

One of the clearest and most radical redefinitions of the aims of ecology comes from Hengeveld and Walter (1999, further references therein) and Walter and Hengeveld (2000, further references therein; see also Walter 1995 and Walter and Patterson 1995). Hengeveld and Walter distinguish two paradigms in current ecology, which have coexisted for some time but are mutually exclusive; they are the demographic and the autecological paradigms. The former, which is the better developed and generally accepted approach, accepts that different species are demographically similar although they fulfill different functions in communities. Intra- and interspecific competition are of paramount importance, leading to coevolution of species by optimization processes. Optimization is thought to be possible because the abiotic component of the environment is on average constant. In contrast, the autecological approach accepts species as dissimilar entities which are affected by abiotic as well as demographic factors. Optimization cannot occur because the environment is very variable in space and time. In the demographic paradigm, the important question is why do so many species share the same resources, and the emphasis is on evolution as the result of short-term ecological optimisation processes. In the autecological paradigm, the central question relates to how species arose and how they persist within a variable and heterogeneous environment. It focuses on the idiosyncratic nature of adaptations, species and their spatial response to environmental circumstances. In the demographic paradigm, nature is balanced, i.e., there are population equilibria maintained demographically by biotic processes; structured communities exist that consist of populations of several species and are saturated with species that optimally partition resources; and ecological-evolutionary processes occur in a discrete locality. In the autecological paradigm, because of the continually changing environmental (biotic and abiotic) conditions which do not permit optimization, emphasis is on survival and reproduction and not on quantification and comparison of differences in reproductive outputs between species. Physiological, morphological and behavioral traits of individual organisms which are important in ecology and evolution deserve attention. In the autecologist's view, populations are temporary and dynamic aggregations of individuals in a stochastically fluctuating environment, they are not abstractions of ecologists in the sense used by demographic ecologists. Autoecology, therefore, is not local like demographic ecology, and its focus is not on balancing numbers across populations, as in metapopulation ecology.

Clearly, Ramensky, Gleason, Andrewartha, and Birch, among others, are autecologists in the sense of Hengeveld and Walter, whereas the majority of ecologists past and present have been demographic ecologists.

Of very great importance are the studies by Brown, Gillooly, Allen and collaborators. Their aim is to explain patterns by the first principles of body size, temperature, and stoichiometry (Allen *et al.* 2002; Gillooly *et al.* 2002; Brown *et al.* 2004). This "metabolic theory of ecology" does not rely on equilibrium assumptions, and can be expected to become increasingly important in the future (see pp. 184–185).

The approach of NKS ("New Kind of Science", Wolfram 2002), and in particular the use of cellular automata as applied by Wolfram (1986, 2002), is also potentially very significant. The basic idea of NKS is to run many computer programs based on different "rules" and to see how they behave. A cellular automaton consists of rows of cells; each cell has a state associated with it, for example, either black or white. Arbitrary rules specify how the automaton develops, i.e., how a cell evolves from one computational stage to the next, based on its previous state and that of its neighbors. Extensive studies have shown that very simple rules (as measured by the number of instructions) lead to simple repetitive patterns, but that a slight increase in a rule's complexity may lead to very complex, apparently random (or better pseudo-random) patterns. Increasing the complexity of rules above a certain threshold does not lead to a further increase in the complexity of patterns.

Other systems, such as mobile automata, tag systems, cyclic tag systems, Turing machines, substitution systems, sequential substitution systems, register machines and symbolic systems such as Mathematica™ (for definitions see Wolfram 2002) follow the same principle, i.e., complex patterns can be created by rules whose complexity lies between a lower and upper threshold.

Traditional science usually considers systems that satisfy certain constraints. Cellular automata can be adapted to the use of such constraints. For example, a constraint can specify that every white cell should have a certain kind of neighbor. Many of such rules lead to simple repetitive or nested patterns, but some lead to complex and pseudo-random patterns.

The Principle of Computational Equivalence is of foremost importance in NKS; it states that a pattern corresponds to a computation of equivalent complexity, in all but the most simple systems. In other words, the computations necessary to predict the fate of any complex system require at least as many steps as are contained in the system itself. This implies that, for most systems, concise predictive laws cannot be found, i.e., they are irreducible, which implies that there is a fundamental limit to traditional science. In other words, it is impossible to predict by some mathematical equation, i.e., a concise "law", what the pattern some steps down will be. This can only be predicted by computations based on the rule. There is no possibility of short-cutting the process of computation.

Not all scientists agree with Wolfram's claim that cellular automata modelling has led to a scientific revolution (e.g., Molofsky and Bever 2004 for ecology), but some studies have already used cellular automata successfully for investigating specialized ecological patterns (see Molofsky and Bever 2004 for examples). Here, I briefly outline Wolfram's (2002) application of NKS to evolution.

The evolutionary process can be interpreted as random searches for programs carried out in order to maximise fitness. If rules are simple, iterative random searches may find the best solutions relatively quickly, but if rules become even slightly more complex, an astronomical number of steps are necessary to even approach an optimal solution. Hence, most species ("programs") are unlikely to be optimally adapted to their niche, they are trapped in suboptimal niches that were easy to find (Wolfram 2002). The numbers of mutations in the course of evolution have been huge, and because relatively simple rules may lead to complexity (pp. 185–186), it is inevitable that some mutations have led to complex patterns early in evolutionary history. There is indeed evidence for this: many complex organisms are very ancient (see, for example, the Aspidogastrea discussed in Chapter 10) and the degree of complexity in many phylogenetic lines has hardly changed, or not at all, over hundreds of millions of years (see, for instance, living fossils such as *Tridacna* and crossopterygians), or organisms may even have become less complex (see, for example, the reduction in the number of headbones during the evolution of higher vertebrates from fishes). Also, complexity of similar features in closely related species, such as pigmentation patterns in cone shells, may vary enormously between species, suggesting that single or few mutations are responsible for the differences, which is possible only in relatively short programs. Furthermore, such patterns closely correspond to patterns generated by randomly chosen cellular automata with simple rules (also suggesting that genetic programs for the patterns are short). Finally, the pigmented shell of certain species is covered (i.e., at least partially obscured) by living tissue, making it unlikely that natural selection is responsible for the evolution of pigmentation patterns.

The relative frequency of simple organisms, and the existence of organs that have some reasonably simple components, may be explained by the assumption that only simple features on which selection can act, can be optimised, and have therefore survived. *In toto*, natural selection does not lead to complexity but tends to avoid it. The fact that evolution has led to many complex organisms, in addition to many simple ones, is a consequence of the random addition of more and more "programs", many of which happen to lead to complex features.

It is interesting to compare the findings of Wolfram with the conclusions of other authors. For example, according to Hengeveld and Walter (1999, further references therein), optimisation of ecological traits can rarely be achieved because environmental conditions are very variable in space and time (see pp. 10–11) and Chapter 10. Wolfram's solution is more radical, since it suggests that optimisation cannot occur even when environmental conditions are constant. Kauffman (1993) doubted the overriding importance of natural selection and, like Wolfram, concluded that many traits of organisms have evolved not because of, but in spite of natural selection. He refers to ideas of the rational morphologists Goethe, Cuvier, and St. Hilaire, who tried to find some logic or laws that "explained similar organisms as variations on some simple mechanisms that generate living forms." In some respects, this is surprisingly similar to NKS. Kauffman concluded that species perform "adaptive walks" in "rugged fitness landscapes" that lead them to local optima where they become trapped. Like Wolfram, he concluded that global optima can seldom (if ever) be reached, species are not optimally adapted.

Regulation and equilibrium in ecological systems: some experiments and a critical discussion of arguments given in favour of equilibrium

Ecologists may argue about the definition and significance of equilibrium and regulation, but there are well documented cases for regulation of natural populations, i.e., for apparent equilibrium conditions. Populations can be regulated by a single predator, as shown, for example, by Silman *et al.* (2003), who studied a dominant rainforest tree species, the palm

Astrocaryum murumuru at a site in southeastern Peru. In the absence of the white-lipped peccary, a seed predator, the density of the palm increased by a factor of 1.7. Re-introduction of the predator reduced the density of the palm to a level found prior to the disappearance of the predator, and the distribution of seedlings with respect to safe sites reverted to that of 21 years earlier. Predators are also involved in regulating densities of the coral reef fish *Gramma loreto*. Webster (2003) performed manipulative field experiments demonstrating that predator-induced mortality is the primary source of density dependence. Over a three-year period (about two generations), local populations underwent regular annual cycles of abundance, the result of seasonal recruitment patterns, but adult density remained almost constant. The fish live in single species aggregations of about 10 to more than 100 individuals. Movement between adjacent reefs is very rare (about 0.7% of tagged residents emigrated per month). Unmanipulated populations were compared with populations in which recruitment was experimentally increased. All fish in each population were captured and tagged, and demographic rates were monitored. Subsequent recruitment was checked by looking for untagged fish. Emigration was checked by looking for tagged fish within about 10 m of each experimental population. Net immigration and net mortality were measured by recapturing all fish at the end of the experiment, and by comparison with the original census taken at the beginning of the experiment. To check for predator responses to density variations, nine additional pairs of populations were filmed. Population size in the experimental populations was measured over three years. Results were as follows.

Population size decreased in experimentally increased populations due to density-dependent changes in demographic rates. Both emigration rates and per capita mortality were density-dependent. However, differences in mortality between unmanipulated and manipulated populations were five times greater than those in emigration. In contrast, both natural recruitment and immigration were density-independent. The experiments thus showed that density-dependent changes were primarily due to mortality. Cause of mortality was neither interspecific competition (other species were not present) nor intraspecific competition (aggression did not increase, and growth rates did not decrease at higher densities), it was predation. Predatory fish spent significantly more time in populations with experimentally increased densities. The overall result was that control and manipulated populations converged completely, becoming indistinguishable.

May and Anderson (1978) have shown that parasites can regulate growth of host populations as well, even in the absence of predation or intraspecific competition (for further references and discussion see Combes 2001). In the context of parasite biology, Combes (2001) defines regulation as maintaining a population by density-dependent mechanisms below the theoretical carrying capacity. Combes further points out that parasite populations are characterized by regulation – where it exists – that occurs independently in each infrapopulation. Exceptions to this are parasites that have long-lived, free-living stages. Regulation in such cases may not occur in the parasitic but in the free-living stages. Furthermore, in strongly aggregated parasites, earlier regulation occurs in the more heavily infected host individuals. Referring to Price (1980) who stressed that the frequencies and intensities of infections with many parasites are so small that they must be considered nonsaturated and nonregulated, Combes counters by claiming that regulation may often not occur at the time of observation, but only when population densities, which fluctuate cyclically, reach a peak. Also, the hosts observed may not be the main hosts. Combes finds support in Keymer (1982), according to whom temporal stability is common in parasite populations, indicative of regulatory processes. Density-dependent regulation is deemed necessary in many cases because otherwise densities would become too great, either by renewed infections or multiplication in the host. Exceptions could be populations in which there is no multiplication in the host and acquisition of new parasites is low. Combes makes the following point, "it is difficult to reconcile the following two statements with the third: (a) parasite populations are almost always aggregated; (b) aggregation favours the appearance of regulation; and (c) parasite populations are usually not regulated." Thus, (c) cannot be true when (a) and (b) are true. Combes therefore concludes that numerous parasite populations are regulated in their hosts. Poulin (1998) also noted that because aggregation is the rule, density-dependent regulation can be expected. However, these arguments are not convincing, because even if populations are aggregated and even if aggregation favours regulation, effective regulation may require much greater population densities in the most heavily infected host individuals than is often observed. Moreover, regulation may be sporadic and thus influence a relatively small part of the population. All this, however, does not mean that regulation in parasite populations does not occur. There are indeed well documented cases of such regulation.

Esch and Fernandez (1993) give examples of density-dependent and density-independent processes in parasites, and Combes (2001) distinguishes three types of density-dependent mechanisms for parasites: (a) decision-dependent regulation (infective stages avoid hosts that are already infected); (b) competition-dependent regulation (population size is limited by limiting resources or some active elimination processes); and (c) host death-dependent regulation (most heavily infected host individuals die). Combes (references therein) gives many examples for all three mechanisms. Thus, certain parasitoids do not oviposit in hosts already infected (a). Several intestinal trematodes and cestodes were shown to have reduced fecundity at high intensities of infection (b), and density-dependent regulation may be host-mediated (certain birds heavily infected with trematodes are more easily preyed upon by predators) (c). The last two mechanisms are probably the most significant.

Concerning the first point, Combes gives the example of the wasp *Venturia canescens*, which is a parasitoid of the moth *Ephestia kuehniella*. The wasp can (but does not always) decide not to oviposit in moths already infected.

Concerning the second mechanism, Combes (2001) and Esch and Fernandez (1993) describe in some detail the example of the cestode *Hymenolepis diminuta*, which has been well investigated by several authors. Final hosts are rats that become infected by eating intermediate beetle hosts containing larval cysticercoids. In various experiments, different dosages (1–20) of cysticercoids were applied. Two months after infection, the length, weight, and egg production of each tapeworm were measured, and found to be inversely proportional to the number of cysticercoids given. In fact, the greatest number of eggs produced per rat was found in those rats that had received a single cysticercoid.

Concerning the third point (host-mediated regulation) mentioned by Combes, this may be brought about in several ways. For example, heavily infected host individuals may be weakened and either die as a result of the infection, or are more likely to be eaten by predators, The latter has been shown for red grouse in Scotland infected with the nematode *Trichostrongylus tenuis*. Those grouse killed by predators were shown to be heavily infected. Because of the highly aggregated distribution of most parasites in their host populations, such selective disappearance of the most heavily infected host individuals may have very significant effects on the parasite population as a whole.

Bradley (1974), also contradicting Price (see above), suggested that parasite populations are controlled by low colonization probabilities only in unfavourable habitats; in more favourable ones they are controlled by factors such as reduction in the number of parasites that can develop, a view supported by studies on immune responses of human and domestic hosts to parasites. However, the latter systems are not natural in the sense of representing naturally occurring parasite population sizes. Furthermore, the important point made by authors most critical of the importance of equilibria and competitive effects is that most habitats are unfavourable, and that our views have been distorted because there is a tendency to select systems for study that occur under favourable conditions (e.g., snail populations that are infected with a diverse trematode fauna, and not those that contain none or few trematodes and are too boring to be studied (see pp. 131–134). More generally, species, populations, and sites studied are not selected at random, but results from these studies are nevertheless treated as representative.

Nonequilibrium in populations and metapopulations: some empirical studies

In the previous section I discussed experimental evidence for equilibrium conditions, in this section I discuss some examples that suggest nonequilibrium.

As mentioned earlier, Cappucino (1995) defined nonequilibrium in a very narrow way, claiming that only randomly walking populations should be considered unambiguously nonequilibrial. Time to extinction of random-walking populations may be very long (Nisbet and Gurney 1982; Murdoch 1994), or almost infinite (Middleton 1993; Murdoch 1994). But even in populations that are not random-walking, i.e., that reach equilibrium when not disturbed, apparent nonequilibrium after disturbance may last for long periods. Alternatively, disturbances may be so frequent that even in fast-recovering populations equilibrium may seldom be obtained. For many species, nonequilibrium conditions resulting from disturbances are perhaps so long-lasting that equilibrium will never be established (Murdoch 1994). I will now discuss some examples; some further, more detailed, examples are discussed in Chapter 7.

A very well documented and analysed example of persistent instability in an animal population is that of the Soay sheep, *Ovis aries*, on Hirta in the St. Kilda archipelago, discussed by Clutton-Brock *et al.* (1991; see also Clutton-Brock and Pemberton 2004). The sheep were originally introduced into the archipelago from Europe. After evacuation of the human population, 107 sheep were introduced into Hirta, the largest of the islands (638 ha), in 1932. Numbers increased rapidly (to about 650–700 by 1948). Between 1955 and 1990, sheep were counted each summer, but the analyses discussed in Clutton-Brock are restricted to 1961–67 and 1985–90. Between 1959 and 1968, most lambs in a certain area were caught, weighed, sexed, and tagged within the first month of life. During both periods selected for the study, population size varied from about 600 to nearly 1600, as a result of high over-winter mortality. Mortality was particularly pronounced among lambs and rams, and was caused mainly by starvation. There were four population crashes, but there is evidence for additional crashes during the periods not included in the analyses. Winter mortality increased with population density in a non-linear fashion. No density dependence was found for neonatal survival or overall fecundity. The authors conclude that the persistent instability is probably caused by the consistently high fecundity due to the super-abundance of food in summer. Consequently, population size rises above carrying capacity of the island, leading to the crashes.

In summary, the results of this study show the frequent occurrence of nonequilibrium conditions (i.e., a marked over- or under-saturation). Indeed, periods during which the population was close to an equilibrium point were extremely short, although the overall result was "regulation": the population fluctuated between certain minima and maxima.

Many relevant examples of often strong and irregular fluctuations in population sizes of plants and animals, and the factors that may be responsible, can be found in Hassell and May (1990), who organized and edited a Discussion on "Population, regulation and dynamics." I discuss two of the contributions from that Discussion, one on fish, the other on plant populations.

Shepherd and Cushing (1990) drew attention to a large body of evidence for long-term changes in the sizes of marine fish populations. They discuss long-term changes in recruitment of North Sea sole, North Sea plaice, Iceland cod, northeast Arctic cod, and North Sea herring in detail. All these species are of great economic importance and have been studied over many years. Fluctuations are great and appear to be quite irregular, but the causes are not known, although climatic changes may be partly responsible. The authors give the following circumstantial evidence for the possible involvement of regulatory processes:

(1) fish populations are known to sustain levels of fishing mortality several times greater than levels of natural mortality (possible evidence for strong regulation);
(2) fish stocks have persisted for centuries with few extinctions or explosions (possible evidence for regulation, but regulation is not necessarily strong); and

(3) there are high levels of fluctuation in recruitment (possible evidence for weak regulation, except in the earliest life-cycle stages).

The authors point out that the time until effective explosion or extinction may be long (possibly 100 years for thousandfold changes in abundance) under weak regulation, but that there are few records implying greater persistence of particular fish stocks than this. Furthermore, fluctuations in population size are so great that, in spite of the circumstantial evidence for regulation, analysis of stock recruitment diagrams rarely provides evidence for or against regulation. Because of the difficulties in demonstrating density-dependent regulation, the authors performed a simulation study to test whether "stochastic regulation" is capable of maintaining fish stocks over a wide range of fishing mortality rates. The following assumptions are the basis for the presumed stochastic regulation:

(1) recruitment of fish stocks is highly variable but non-negative, leading to a skewed probability distribution better described, for example, by a log-normal than a symmetric (such as a normal) distribution;
(2) variability of recruitment increases with declining stock sizes; and
(3) median recruitment is deterministically determined.

However, assumptions (2) and (3) have not been proven. Nevertheless, the authors conclude that it seems indeed plausible or even likely that the only regulatory mechanism for fish stocks is stochastic, i.e., an increased and non-normal variability in low stock sizes. The effect would be "strong regulation in the mean," resulting from the increasing excess of the mean over the median when population sizes are small. These results may be suggestive, but require confirmation by an analysis of real long-term data sets. Is stochastic "regulation," even if it occurs, really strong enough to produce the effects shown in the simulation study? Also, is the term regulation appropriate, considering that "regulation" in the simulations is based on purely stochastic events, and that density-dependent processes are not involved? Finally, as pointed out by the authors themselves, an approach at least as valuable would be to examine empirically the causes of variability themselves.

Crawley (1990) reviewed long-term studies of plant populations and found evidence for regular patterns in relatively few of these; density-dependent processes are important in certain populations for which several generations have been monitored, and, in contrast to animal populations, where the significance of competition is "a matter of debate," it is "quite clearly of over-riding importance" in most plant assemblages.

Crawley distinguished three categories of plant populations. Only one of these, comprising a relatively small number of species, shows "typical" population dynamics at a given point in space, following more or less predictable trajectories described by the $N_{t+1} = f(N_t)$ model. The second category includes a large number of short-lived plants with ephemeral, pulsed dynamics lasting a single generation, recruitment almost entirely determined by germination biology, and frequency and intensity of disturbances. The third category includes some long-lived plants for which population patterns cannot be established because of the long life spans. Furthermore, most plant species are successional, whose recruitment depends on the death (either resulting from senescence or disturbances) of dominant plants.

An example discussed by Crawley with strong evidence for the effect of disturbance but no evidence for equilibrium is a study by Hubbell and Foster of plant species on a permanent plot on Barro Coloradi island. They compared the flora in 1980 and 1985. In 40% of the plant species, abundance had changed by more than 10%. There had been "catastrophic" mortality during a drought year, and rare species declined more strongly and common species increased more strongly than expected by chance alone. No evidence was found that the plant assemblage was in equilibrium.

A discussion of tropical rainforests in Chapter 8 addresses the question of the relative significance of nonequilibrium and equilibrium in plant assemblages in greater detail.

Concerning the effects of disturbances on metapopulations, Hanski (1999, references therein) reviews work on the Glanville fritillary butterfly, *Melitaia cinxia*, on southwestern Finnish islands. The work is one of the few thorough long-term studies and is therefore discussed in somewhat greater detail. Habitat reduction led to nonequilibrium conditions and projections showed that a new equilibrium would only be established after many years.

The distribution of the butterfly in Europe has markedly declined over the last decades, and it became extinct on the Finnish mainland in the late 1970s. It still survives on the Åland Islands, southwestern Finland. Hanski and Kuussaari (1995) and Hanski (1999, references therein) describe in great detail the metapopulation dynamics of the butterfly on the Åland Islands with about 1600 dry meadows suitable for colonization, beginning in 1991. Examined were the four conditions necessary for persistence at the metapopulation level. These conditions are: (1) the species has local breeding populations in relatively discrete habitat patches, with some migration between patches but most individuals interacting with individuals in the same patch; (2) single populations are too small to have a long lifetime relative to that of the metapopulation; (3) the patches are close enough to permit recolonization; and (4) local dynamics are insufficiently synchronized for a likely occurrence of simultaneous extinction of all local populations (see also Hanski and Gilpin 1991).

Larvae of the species have two host plants in the area. Females lay eggs in groups of up to 200. In 1993–97 there were 300–500 local populations, some very small but most with 50–100 larvae in late summer; some 30–50% of larval groups in the larger populations were estimated to have not yet been discovered. Habitat patches containing the local populations are well delimited by the surrounding environment. Estimates showed that about 80% of the butterflies spent their entire life in the natal patch. Even the largest population of about 650 butterflies became extinct after a few years. Mark–recapture experiments showed that about 9% of 741 recaptures were from a new patch. About 15% of males and 30% of females moved to another patch during their lifetime. Patch area, density of flowers and open patch boundaries determined emigration and immigration rates. In a study of 1737 marked butterflies, most migrating butterflies moved only a few hundred (less than 500) m, although up to 3.1 km migration distance was recorded. Among 906 colonizations observed between 1994 and 2000, the mean distance from the nearest population was 0.6 km and the longest 6.8 km (van Nouhuys and Hanski 2002). The annual colonization rate of empty patches was approximately 15%.

An intensive study from 1991 to 1993 demonstrated largely asynchronous local dynamics. Asynchrony appears to be maintained by an interaction of weather and habitat quality, and the actions of two hymenopteran parasitoids, one of which also has distinct metapopulation structure. One of the parasitoids (which have several hyperparasitoids) caused larval mortality from 0–100%, varying between populations and from year to year, indicating a very great spatial and temporal variance. This means that despite possible density- dependent regulatory mechanisms, the mostly very small local populations and even the largest populations, have a high risk of extinction. Thus, between 1993 and 1995 more than 400 population extinctions were recorded. There are many suitable but empty habitats (meadows). Application of the incidence function model (IFM) showed that fluctuations in the proportion of occupied patches over time were the result of spatially uncorrelated stochasticity in extinctions and colonizations (although in real systems environmental

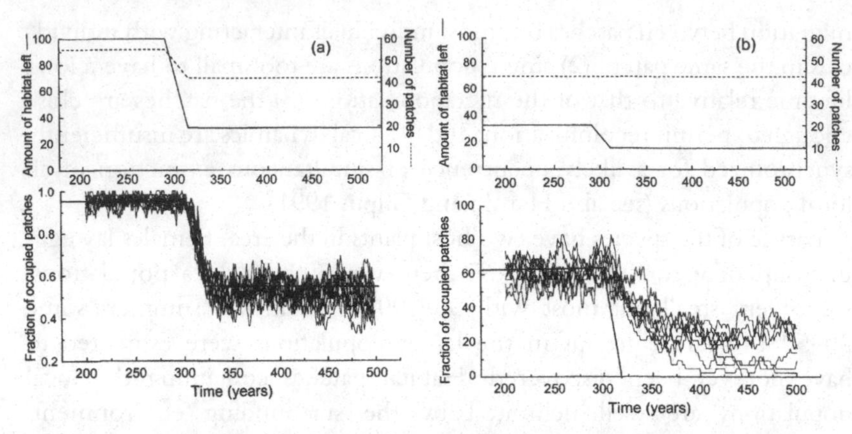

Figure 1.1. Metapopulation dynamics of the Glanville fritillary butterfly. Size of the metapopulation measured as a fraction of the occupied patches *P*. The dotted line in the upper panels gives the number of distinct patches; the continuous line gives the pooled area of these patches. Panel (a) (lower) gives the predicted equilibrium metapopulation size (thick line) with ten replicate predicted trajectories before, during, and following an observed reduction in habitat area over a 20-year period (shown in upper plot). Panel (b) shows similar results, but there was a further 50% reduction in area of each of the remaining patches (upper plot); in this case, the equilibrium moves to metapopulation extinction although a substantial amount of habitat remains. The simulated trajectories show a slow (over many years) decline with much variation. Note the delay in metapopulation dynamics, as indicated by the difference between *P* (the fraction of occupied patches), and the calculated equilibrium values given in the thick line. From Hanski (1999). Reprinted by permission of Oxford University Press and the author.

stochasticity is spatially often correlated). Results show that there are alternative (more than two) equilibria. Over longer time scales, often due to changes in landscape structure which may occur on or around the same time scale as extinctions and recolonizations, metapopulations may not be at stochastic equilibrium with landscape conditions but are tracking the changes with some delay. Such nonequilibrium conditions are likely for the fritillary populations in a 25 km^2 area in northern Åland. Projections demonstrated that, after habitat reduction, a new equilibrium would be reached after many (tens or even hundreds of) years (Figure 1.1).

Generally, "transient time," i.e., the time between disturbance and re-establishment of equilibrium, is longest when the disturbance is large, when a species is close to the extinction threshold and has a slow turnover, and when there are only few dynamically important habitat patches (Ovaskainen and Hanski 2002).

As shown above, the effects of environmental disruptions on population dynamics are clear, but it should be emphasized that nonequilibrial

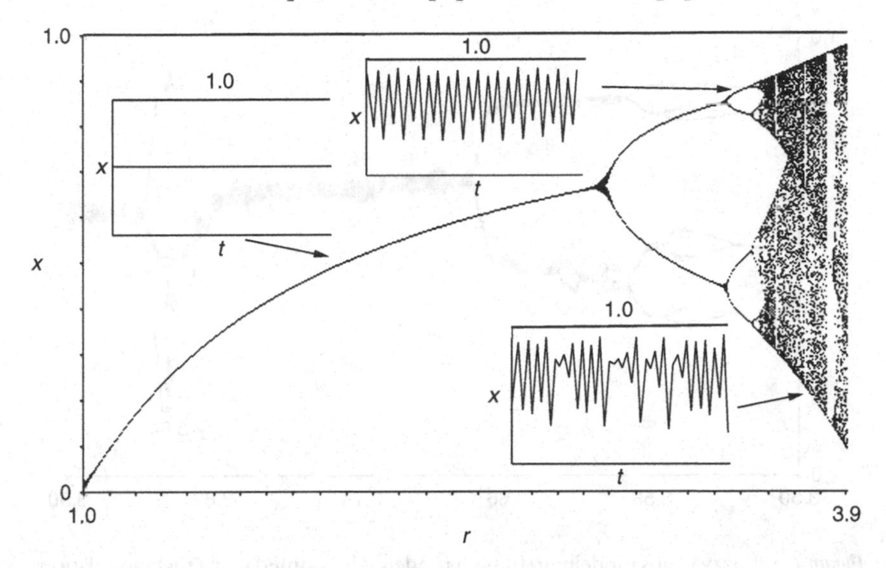

Figure 1.2. Bifurcation diagram showing population size *x* (as the proportion of the carrying capacity *1*) plotted against reproductive rate *r*. Insets show change of population size of populations with selected reproductive rates plotted against time *t*. Note: the population size is at a stable equilibrium with single values of *r* until a certain value of *r* (larger than 3) is reached. Now there are two equilibria, and at even higher values of *r* there are 4, 8, and 16. At $r = 3.57$ fluctuations in population size become chaotic. Note: *x* never reaches carrying capacity.

conditions may well arise without the involvement of such disturbances. May (e.g., 1975; and others), in a series of brilliant papers, have demonstrated that chaotic fluctuations in population size simply arise as the result of high intrinsic rates of population growth *r* (Figure 1.2). At low rates, populations are in equilibrium with a single value of population size, when values exceed 3, there are at first 2, then 4, 8, and 16 values, until at about $r = 3.57$ fluctuations become chaotic, that is, apparently random but in fact strictly deterministic. Also, fluctuations are extremely sensitive to initial conditions, i.e., very small differences in initial population size lead to very large differences in future population fluctuations, making predictions impossible.

Considering chaos in metapopulations, Rohde and Rohde (2001) used "fuzzy chaos" modelling to show that the degree of chaos is reduced when subpopulations composing a metapopulation and distinguished by different reproductive rates are largely segregated. In such metapopulations, the width of the chaotic band becomes much narrower, and it

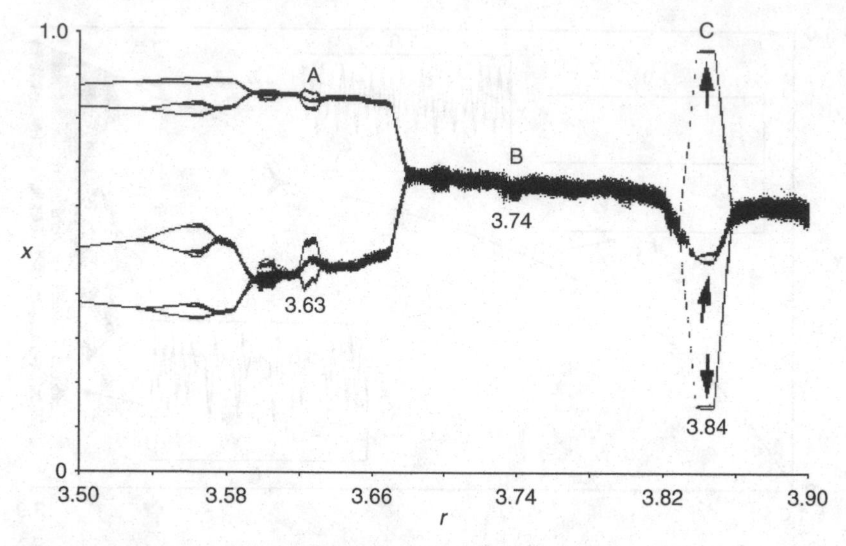

Figure 1.3. Fuzzy chaos modelling. In such models it is assumed that a metapopulation consists of subpopulations differing in reproductive rates. The degree of chaos depends on the number of subpopulations, the range of their reproductive rates, and the initial size of the metapopulation. In the example illustrated here, the initial population size $x_0 = 10^{-2}$, the number of subpopulations = 1000, and the range of reproductive values $r = 10^{-2}$. Note that the chaotic band has contracted compared with Figure 1.2, but there is still chaos and at various values of *r* there are bifurcations indicating more than one widely diverging point of equilibrium.

is the narrower the more subpopulations are there (Figure 1.3). However, chaos still occurs and conditions are largely unpredictable, partly due to the fact that bifurcations with more than one equilibrium point arise at certain values of *r*.

In conclusion, the last two sections have shown that sometimes there is evidence for long-lasting equilibrium conditions (or long-lasting relative constancy in population size) in certain populations. However, severe and apparently irregular fluctuations in population size are common among plants and animals, although reasons for such fluctuations are in many cases little understood. Both in populations and metapopulations, environmental disruptions are important in causing nonequilibrium, although nonequilibrium can arise even in the absence of disturbances.

Defining the problem

We can use the excellent discussion of Rosenzweig (1995) as a starting point for defining the problem discussed in this book. His discussion of steady states over long geological time scales begins with the statement that only negative feedback variables can produce them. The overall dynamics, at least for species within one trophic level, are determined by the following:

(1) speciation rates per species decline with increasing species richness; and
(2) extinction rates per species grow with increasing species richness.

He further concludes that extinction and speciation rates must somehow be balanced because if extinction rates were greater than speciation rates, "we wouldn't be here," and if speciation rates consistently exceeded extinction rates "the Earth would be an ever-expanding collection of species. However, the data say otherwise." Rosenzweig referred to Whittaker (1972) who had claimed that the steady state is eliminated if more trophic levels are added. According to Whittaker, "Species are niches for other species," leading to a continuing increase of diversity. Contradicting Whittaker, Rosenzweig argues that even if positive feedback mechanisms exist in one part of the system, the system as a whole does not necessarily "get out of control" because the increasing numbers of species must be supported by the same resource base. Rosenzweig does not rule out that equilibrium dynamics may exhibit complicated, nonlinear behaviors that can be expressed, for instance, in limit-cycle oscillations or almost-periodic trajectories. He also admits that diversity grew over hundreds of millions of years through colonization of new habitats, such as the muddy ocean bottom, perhaps by an increase in "versatility," and possibly some other factors, for which, however, there is no definitive evidence.

Concerning Rosenzweig's assertion that "data say otherwise," i.e., that there has been no significant increase in diversity over geological time, except for that resulting from colonization of large new habitats and some increase in "versatility," the evaluations of fossil evidence by Benton (1995, 1998), Jablonski (1999) and Jablonski *et al.* (2003) show a very marked increase in diversity over evolutionary time. Even if the evaluation of Courtillot and Gaudemer (1996) is accepted, who claimed that a steady state has been reached several times in geologic history, the fact still remains that after each "steady state" a jump to much higher diversity occurred. Furthermore, there is no reason whatsoever to assume that just now in history a point of no further increase has been reached (excepting the possiblity of man-induced mass extinctions). In other words, fossil evidence alone would suggest that the Earth is indeed an ever-expanding collection of species.

Concerning the other point made by Rosenzweig, that the system as a whole cannot get out of control even if species at other trophic levels are added, because it must be supported by the same resource base, it has been shown by several authors that vacant niches exist in various systems which lend themselves to colonization even if the resource base as a whole does not expand. In other words, resources are not exhausted. Moreover, why should there not be an increase in diversity by subdivision of niches? It is the aim of the following chapters to provide detailed support for these assertions, and to evaluate the relative importance of equilibrium and nonequilibrium by providing evidence suggesting equilibrium and nonequilibrium conditions over evolutionary time and in extant ecological systems.

2 · *Nonequilibrium in communities*

Definition and evolution of communities

Ecological community is not a term that is used uniformly by all authors, and there is much disagreement about how communities may have evolved. In this section, I discuss definitions given for ecological communities and their evolution.

Giller and Gee (1987) review the different ways the term community is used, and the problems arising from these different usages. Fauth *et al.* (1996) defined communities as all species co-occurring at the same time, irrespective of taxon; guilds are species that use the same class of resources. They defined assemblages as groups of species of one taxon (e.g., birds) within a community. According to Cornell (2001), communities are collections of species living contemporaneously in the same place, consisting of individuals that are spatially interspersed, with the potential of direct or indirect interaction. Following Whittaker, Levin (1992) points out that "communities" and "ecosystems" are arbitrary subdivisions of a "gradation of local assemblages." Communities are not well integrated units, because species within them respond individualistically to the environment. Lawton (2000) asks: how many species constitute a community? There is no logical break between populations of single species and of many, and there is a tendency to take several species of a single taxon as comprising a community. Lawton points out that entire communities are almost impossible to study, with the exception of some in very simple habitats such as water-filled tree holes. However, there are many other habitats that are well defined and have a characteristic set of species which are not found elsewhere. Such species sets comprise easily defined communities. Many of them are quite simple and can easily be studied. Examples are parasite communities on the gills of fish and in the alimentary tract of various vertebrates.

The central question of community ecology is whether communities in a habitat comprise all species that have happened to arrive there, or only a subset of those species with characteristics that permit their coexistence (Roughgarden 1989). Elton (1933) subscribed to the latter view, whereas Gleason (1926) subscribed to the former. (For a discussion of the concept of community see also Brown 1995, further references therein.) Comparing the individualism of species composing communities advocated by Gleason, Whittaker and others, with the structured approach of MacArthur, Hutchinson, and others, Brown emphasizes that both approaches are not incompatible. He points out that it is not correct to equate Gleason's individualism with the influence of abiotic conditions, and structure with the effects of biotic interactions. Both interactions and reactions to abiotic factors are individualistic, depending on unique features of the species; and the environment can impose patterns of community structure, as indicated by convergent structural and functional features of only distantly related organisms in different continents. Brown discusses some examples.

There has been much discussion about how communities should be defined, on the basis of palaeontological evidence (see the controversy of Anderson 1995 and Walter and Patterson 1994, 1995). Anderson stresses that interactions between species in addition to their co-occurrence, define a community. The latter authors come to the conclusion that palaeoecological evidence supports the view that species have associated individualistically and have not co-adapted to other species in the extant community, which – however – does not mean that they have evolved independently of other species. They point out that, in order to demonstrate co-adaptation, conclusive evidence for resource partitioning has to be given, and that no studies of purported cases of local character displacement have satisfied the minimum criteria outlined by Connell (1980). Emphasis should instead be placed on the study of individuals and their adaptations, and a community should be viewed as a level of organization and not as an entity. One looks for repeated patterns and such patterns occur most consistently where adaptive mechanisms exist. Species that form extant communities have evolutionary histories that are independent of other species in extant communities, which – however – does not imply that the species have evolved in a vacuum. Lawton (1984a) considers that, with regard to herbivore communities, coevolution should be ignored as an explanation "until the data absolutely force us to do otherwise." And, based on his study of insects of bracken fern, "Chance may have played a large part in determining which species colonized the plant over evolutionary time." Bracken insects use their own locations on the plant and their own mode of feeding, and are

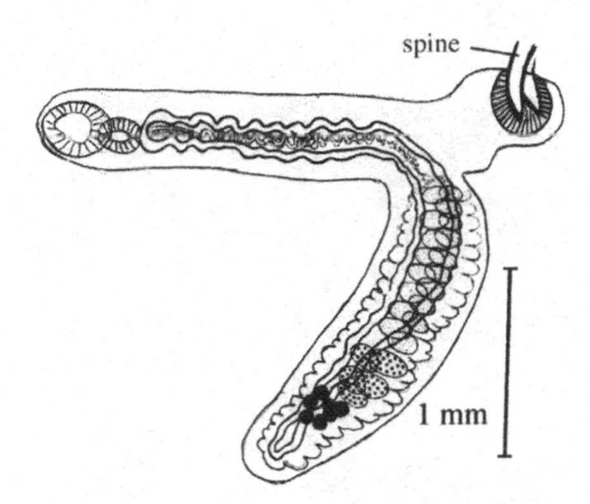

Figure 2.1. The trematode *Syncoelium filiferum* from the gill arches of the teleost fish *Seriolella brama* in New Zealand. It is one of the very few (if not the only) species of trematode infecting the gills of fish, most others occurring in the digestive tract and associated organs. It is attached by means of a very large ventral sucker to the spines of the bony gill arches. From Rohde, Roubal, and Hewitt (1980). Reprinted by permission of the Royal Society of New Zealand.

"insensitive to the presence or absence of other species." With regard to interspecific competition, patterns found in many communities may be explained by competition, but also by alternative hypotheses.

The following example of a very simple community of gill parasites shows the difficulties or, in the absence of fossil data, even the impossibility of making a decision as to whether the individualistic or structured hypothesis is correct. It also shows that the possibility should not be disregarded that a community may be composed partly of individualistically assembled and partly of interacting species. Nineteen *Seriolella brama*, a small to medium sized teleost fish, were examined at a locality in New Zealand (Rohde *et al.* 1980). The gill parasite community consisted of three species, the digenean trematode *Syncoelium filiferum* (Figure 2.1), and the monogeneans *Eurysorchis australis* (Figure 2.2) and *Neogrubea seriolellae* (Figure 2.3). The first species is an opportunistic parasite which infects a wide range of marine teleosts, whereas the other two are, on present knowledge, specific to two species of *Seriolella*. A prerequisite for successful infection with *Syncoelium* is presence of spines on the bony gill arches to which it can attach by means of its ventral sucker. *Eurysorchis* inhabits the bony gill arches, pseudobranchs and mouth cavity, attached to these flat surfaces by means of its large flattened clamps which act as suckers. *Neogrubea* is always attached by means of its clamps to the gill filaments, which are "grasped" by the two valves of the clamps. All three species have clear morphological adaptations to their microhabitats

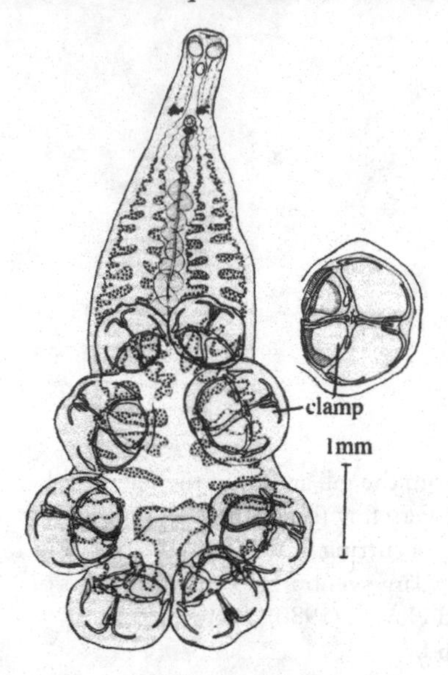

Figure 2.2. The polyopisthocotylean monogenean *Eurysorchis australis* from the gill arches, pseudobranchs and mouth cavity of *Seriolella brama* in New Zealand. Most species of this group have clamps consisting of two valves that can grasp gill filaments (see Figure 2.3), but this species is attached by means of flat sucker-like clamps to flat surfaces. From Rohde, Roubal, and Hewitt (1980). Reprinted by permission of the Royal Society of New Zealand.

and cannot live elsewhere. It is unlikely that any of the species interact with the others because of their spatial segregation, but it cannot be excluded that *Eurysorchis australis*, which has a sucker apparently secondarily modified from a clamp-like one as in *Neogrubea seriolellae*, has been forced into a new microhabitat away from the gill filaments by competition in the past ("the ghost of competition past", Connell 1980), for which assumption, however, there is no evidence. *Syncoelium*, on the other hand, one of the very few trematodes (if not the only one) infecting the gills and the only one using its sucker for attachment to the spines of the gill arches, must have invaded the gills secondarily in the past, coming from the digestive tract where most trematodes live. There is no reason to assume that at any stage in the evolutionary past it had to compete with other species in the new microhabitat (unless, of course, competing species have become extinct, an assumption without basis). In other words, *Syncoelium* very likely associated individualistically with the gill community of parasites. Also, it cannot be excluded that other species trying to enter the community in the past have been "rejected" because

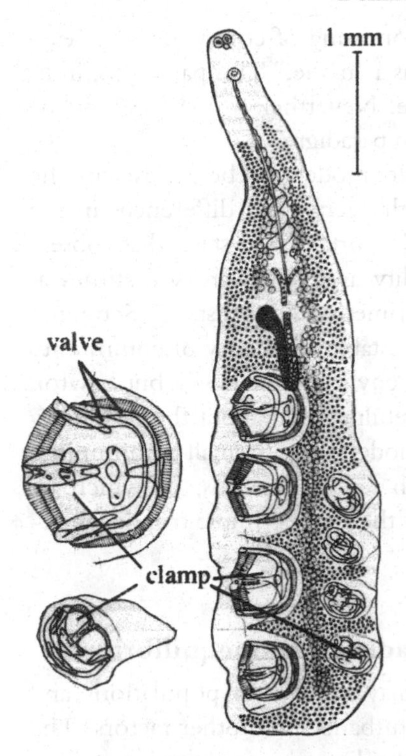

Figure 2.3. The polyopisthocotylean monogenean *Neogrubea seriolellae* from the gill filaments of *Seriolella brama* in New Zealand. The two valves of its clamps close in around gill filaments and "grasp" them. Attachment to flat surfaces, such as the bony gill arches, is impossible This results in strict site specificity of the parasite. From Rohde, Roubal, and Hewitt (1980). Reprinted by permission of the Royal Society of New Zealand.

suitable habitats were already filled, although the existence of many vacant niches on the gills makes such an assumption unlikely. Moreover, the hypothesis cannot be rejected that some of the extant species are newcomers that have competitively replaced others in the past. In other words, the extant community may well be a subset of the species that could have assembled because they are suitable for the habitat. Which means that, in the absence of fossil data, it is impossible to test whether populations in a habitat consist of all species that have happened to arrive there, or only of a subset of those species with characteristics that permit their coexistence (Roughgarden 1989, see above).

Concerning plant communities, according to Palmer (2001), "in caricature, the equilibrium paradigm of community ecology states that plant communities are stable entities consisting of competing species – and that species coexist because each has a "niche." This paradigm, in its extreme, has been dead for some time. Nevertheless, it has yet to be replaced by a credible "non-equilibrium paradigm."

Colwell (1984) doubts that community models can be precise, and he emphasizes the need to acknowledge the very great differences in the biologies of various kinds of organisms. Importantly, he states that "observation, logical inference, and plausibility arguments are sometimes as capable of scientific revelation as experiments and statistics." Schoener (1986a) has made a detailed attempt to establish a theory of community ecology based on six organismic and six environmental axes, but Lawton (1999) expresses scepticism about its usefulness and about the possibility of ever establishing in general a useful model for even small communities comprising say 10 or 20 species, with few exceptions, of which he mentions lake communities, because of the latter's simple trophic structure and few key species.

Equilibrium, and disturbance leading to nonequilibrium

We saw earlier that apparent equilibria may well exist in populations, and that nonequilibria may arise due to disturbances and other factors. The same holds for communities. We begin the discussion with two examples, one providing evidence for apparent equilibrium, the other for nonequilibrium after disturbance. We then discuss some theoretical considerations and provide more experimental evidence, as well as evidence from invasions.

A particularly impressive example of insect populations controlled by a parasitoid, and kept in apparent equilibrium over long periods, was studied by Murdoch and collaborators. Over many years, they studied the California red scale, *Aonidiella aurantii*, an insect pest of citrus, and its biological control agent, the parasitoid wasp *Aphytis melinus* (Murdoch 1994). The wasp controls populations of the scale very effectively. Over decades, densities were less than 1% of those present before the wasp was introduced. In one example, 10-year sampling (20–30 generations of the scale, 50–80 generations of the wasp) showed only minor fluctuations around a mean density that was apparently constant. There was no evidence for local extinction. The constancy of population size contradicts theoretical predictions, according to which, populations suppressed far below their resource limits, should be unstable with major fluctuations. How can this be explained?

Eight potentially stabilizing mechanisms were examined. They were: (1) parasitoid aggregates in response to host density; (2) parasitoid aggregations are independent of host density; (3) there is temporally density-dependent (possibly delayed) parasitism; (4) the parasitoid sex-ratio is density-dependent; (5) host-feeding is density-dependent; (6) predation is density-dependent; (7) there are spatial refuges from parasitism; and (8) metapopulation dynamics. No evidence for any of the mechanisms was found except for (7). However, this mechanism (the existence of spatial refuges from parasitism) was not stabilizing. Additional potentially stabilizing mechanisms based on the responses of individual *Aphytis* to their size-structured host were found in modelling and laboratory studies, but further studies are needed to show whether these or other factors can indeed explain the remarkable constancy.

Good experimental evidence for disturbance leading to nonequilibrium in plant communities comes from the analysis of Tilman (1982), according to whom plant communities at Rothamsted, England, may not have re-established equilibrium conditions 100 years after resources had been disturbed, indicative of the large time lag in tracking the resource change. The Rothamsted Park Grass Experiment is particularly well documented. It began in 1856. The eight-acre pasture, used for 200 years prior to the commencement of the experiment as grazing land, had been divided into 20 plots, two used as controls, the others for various kinds of fertilizer treatments. The first quantitative survey of species composition was made in 1862, and it has been repeated several times since. The evaluation of Tilman showed that there was an increasing trend of dominance of *Aleopecurus* and *Arrhenatherum* in two plots since the early stage of the experiments, but the full effect of fertilizer treatment was not realized for almost 100 years. In one of the plots, a new equilibrium in the plant community may not even have been reached after 100 years. The reason appears to be that most species in the communities are long-lived perennials, often reproducing clonally by means of underground tissues. How long it may take to re-establish equilibrium in plant communities, is indicated by evidence that clones of some plants may reach an age of thousands of years.

How can equilibrium and nonequilibrium be explained? Levin and Paine (1974), in an important paper, promoted an alternative view to the then generally accepted view of communities as systems which are in equilibrium. They view communities as spatial and temporal mosaics of open and integrated patches. Disturbances interrupt the "march to equilibrium," and the whole system must be viewed as in balance between extinctions and the immigration and recolonization abilities of the various species within. Disturbances increase heterogeneity by facilitating local differentiation through random colonization and founder effects, and through interrupting natural successions. For other important discussions see Rosenzweig (1995) and the review by Petraitis *et al.* (1989). Hengeveld (1994) stressed the more general significance of variations in living conditions, "as living conditions vary in time and space, species continually have to adapt spatially and genetically, implying that ecological optimization theories based on equilibrium assumptions do not apply." Harris (1986) demonstrated that environmental disturbances occur with such frequency that competitive exclusion in phytoplankton species does not occur, leading to nonequilibrium. These conditions may explain Hutchinson's "paradox of the plankton," i.e., the fact that many more species using similar resources co-occur than expected in a competitive equilibrial world. Nevertheless, even in phytoplankton there are repeated seasonal patterns, i.e., there is a certain degree of predictability imposed by strong environmental pressures.

A considerable body of evidence suggests that predation allows competing species to coexist (e.g., Caswell 1978 and references therein). In a model by Caswell, a predator initiates new "cells for nonequilibrium growth of the prey species," permitting their long-term coexistence. He further suggests that interactions as represented in his model may be of major importance in the real world.

We use an important paper by DeAngelis and Waterhouse (1987) to illustrate the factors that may be responsible for equilibrium in communities, and those that lead to nonequilibrium. According to DeAngelis and Waterhouse, equilibrium and stability are not sharply defined for real systems. They distinguish stably interactive communities, unstably interactive communities and weakly interactive communities, referring to May (1973) whose "results, in particular, made theoretical ecologists acutely aware that there are real difficulties with the idea of the ecosystem as a balance of interacting species." There are two potentially disruptive influences on ecosystems: (a) nonlinear feedbacks and time-lags in the interactions of biological systems may lead to instability, and (b) fluctuations of environmental factors may cause stochastic disruptions (Figure 2.4). Five hypotheses, illustrated in Figure 2.5, have been proposed to account for stability in spite of these influences. These are for a, (1) certain types of predation, inter- and intraspecific interactions, and other

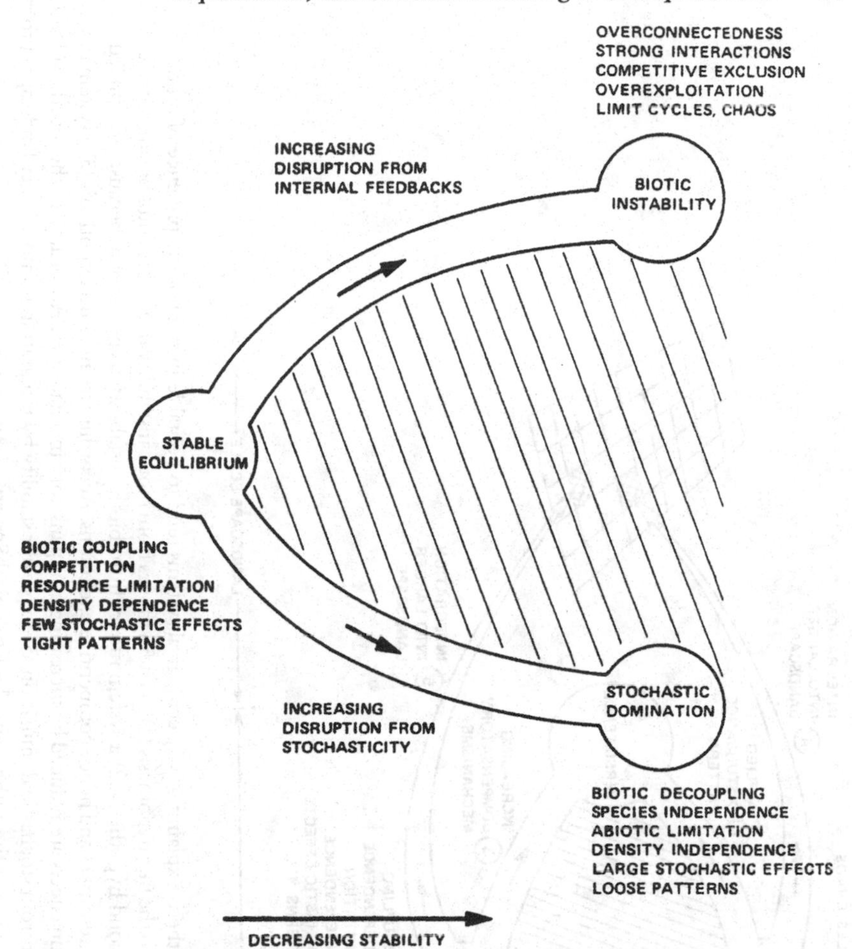

Figure 2.4. Diagram representing ecological systems in a scale from stable (left of the diagram) to instable (right of the diagram). At stable equilibrium there is a high degree of biotic coupling, i.e., species compete for limited resources, their populations are density-dependent, and there are few stochastic effects. Instability can be caused both by biotic instability due to internal feedbacks (e.g., very strong interactions between species leading to overexploitation of resources and competitive exclusion), and by stochastic domination due to strong environmental fluctuations (e.g., species are largely independent, their populations are not controlled by density-dependent processes, and stochastic effects are significant). From DeAngelis and Waterhouse (1987). Reprinted by permission of the authors and the Ecological Society of America.

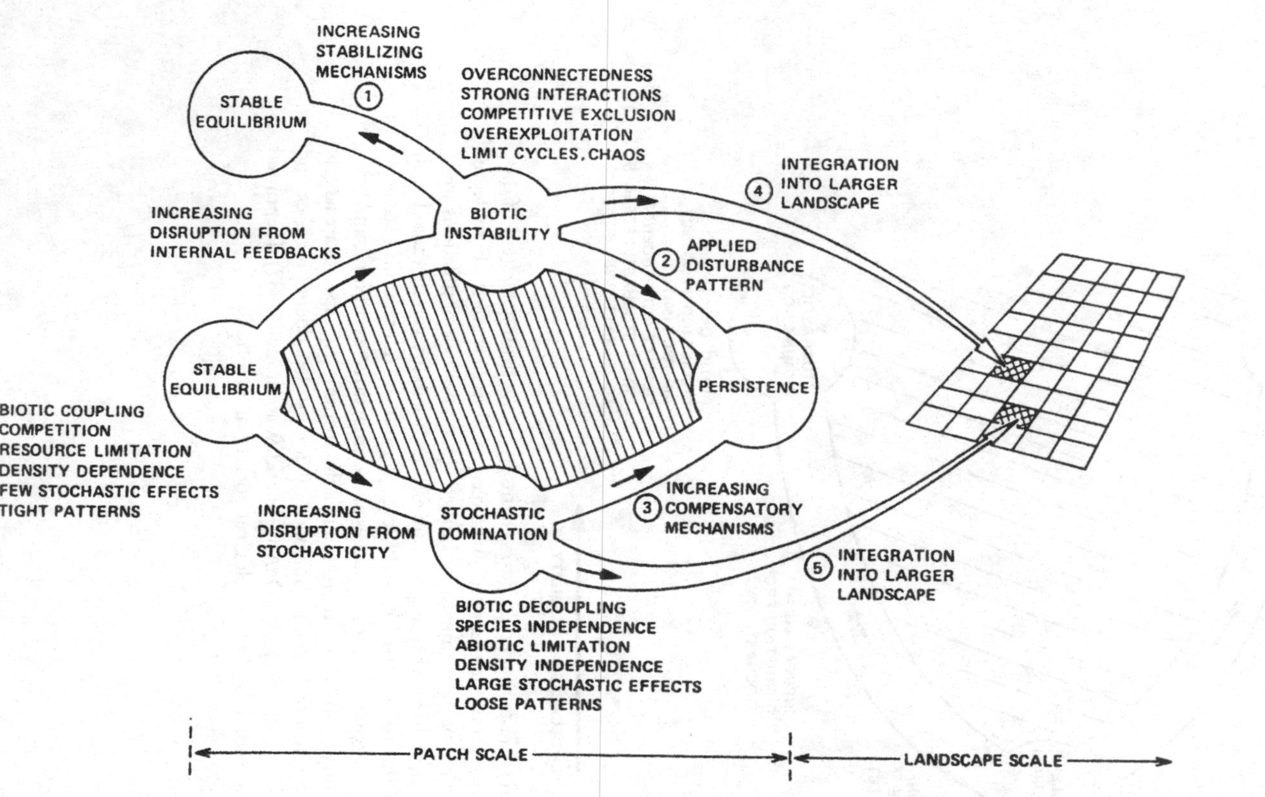

Figure 2.5. Diagram showing five general types of hypothesis explaining why ecological systems tend to be stable in spite of the presence of biotic instabilities and environmental stochasticity. According to the first hypothesis (1) a system moves from biotic instability to a stable equilibrium because of an increase in stabilizing mechanisms. According to the second hypothesis (2) a system moves from biotic instability to persistence because of "applied disturbance patterns" which reduce over-exploitation of resources and hence competitive exclusions. According to the third hypothesis (3) a system moves from stochastic domination to persistence when stochastic effects are reduced by increasing compensatory mechanisms . According to the fourth and fifth hypotheses, a system moves from biotic instability (4) or from stochastic domination (5) to greater stability by integrating it into a larger landscape. From DeAngelis and Waterhouse (1987). Reprinted by permission of the authors and the Ecological Society of America.

biotic factors, act as stabilizers; (2) disturbances interrupt disruptive feedbacks; (4) integration of small-scale-spatial systems into larger ones acts as stabilizer; and they are for b, (3) compensatory mechanisms act at low population densities, ensuring the survival of rare species by increasing the survival rate of the few individuals left (for instance by reduced interspecific competition and increased reproductive effort); and (5) spatial extent and heterogeneity have moderating effects.

Circumstantial evidence for "slow competitive displacement" in tropical rainforest tree communities was found by Hubbell and Foster (1986). Some results may also be explained by equilibrium hypotheses but, overall, the authors conclude that pairwise and predictable interactions between tree species are less important than locally unpredictable disturbances, and influences of biotic uncertainty. Many examples of communities not at competitive equilibrium were given by Harris (1986 and references therein). They include tropical and temperate forests, herbaceous plant communities, lichens, coral reefs, coral reef fishes, and decapod crustaceans. Simulations were used by Moloney and Levin (1996) to evaluate the relative importance of various factors involved in species' responses to disturbances. Simulating population dynamics of three interacting plant species, they considered overall rate of disturbance, size of individual disturbances, and temporal as well as spatial autocorrelation among individual disturbances. Spatial factors (size of individual disturbances and spatial autocorrelation among individual disturbances) played a relatively minor role in the model used. Most important were disturbance rate and temporal autocorrelation in the disturbance regime.

Environmental disturbance can be caused by invasions/introductions of species. Well documented examples are various animals introduced into Australia, including humans, the dingo and feral dogs, foxes, cats, pigs, camels, buffaloes, goats, horses, donkeys, cane toads, and among plants the prickly pear cactus. Introduction of the dingo into Australia many thousands of years ago is thought to have led to the extinction of the Tasmanian tiger on the Australian mainland, humans are known to have exterminated the last Tasmanian tiger in its refuge in Tasmania and many other species in Australia many years ago, and goats, another example, are responsible for large-scale habitat degregation and impoverishment of the flora and fauna. We discuss one example in some detail, that of the European rabbit *Oryctolagus cuniculus*. The rabbit is a native of southwestern Spain, it was spread around the Mediterranean by the Romans, and it was introduced into Victoria, southeastern Australia, in 1862. Over the next 60 years it took over the southern half of Australia, and it is now found in most regions of the country; more than 84% of New South Wales alone is affected. Reflecting the rabbit's origin in the Mediterranean region, populations have the greatest potential for rapid increase in Mediterranean-type temperate zones. In the more arid areas of Australia, population crashes may occur in periods of drought. Rabbits are territorial, and form linear dominance hierarchies. They have native predators such as eagles, and introduced predators such as foxes and cats; up to 70% of the young may be taken by predators. Rabbits are an important pest competing with kangaroos, sheep, and cattle for food resources, although competition between these herbivores is minimal at high biomasses (see pp. 104–108). Consequently, they have an enormous economic impact, causing an annual loss of around A$ 600 million (with which we are not concerned here). But they also cause much ecological damage through habitat degradation and competition (for details see Dickman *et al.* 1993). Rabbits feed on more than 60 plant species, preferring Gramineae and dicotyledons. They prevent regeneration of cassuarine species (*Allocasuarina verticillata*). Coman (1996) recorded 17 new plant species in an area from which rabbits had been excluded, that did not occur in the rabbit-infested control area. Food competition has led to the reduction of native animals, e.g., the Bilby (*Macrotis lagotis*) (Coman 1996), and the Long-billed Corella (*Cacatua tenuirostris*) (Emison *et al.* 1994). But importantly, competition for a limiting resource is not the only factor responsible for species impoverishment. Rabbits may also disrupt the predator–prey cycle and thus cause damage to the native fauna. In the absence of rabbits, predator numbers decline when native prey numbers decline, and it takes some time to rebuild the predator population after the prey population is re-established. In their presence, predators simply switch to the easily available food source, i.e. rabbit, thus maintaining high densities, and maintaining the pressure on native animals as well (for further details and additional references see Nichols 1996, on whom much of this account is based).

All this means is that introduction of a species can lead, at least locally but also on a larger scale, to reduction in diversity and thus to nonequilibrium. A very long (evolutionary) time may be necessary for re-establishment of the original diversity, if it can be re-established at all, i.e., nonequilibrium conditions are very long-lasting indeed.

Our general conclusion for this section is that abiotic and biotic disturbances are important factors leading to nonequilibrium.

Species nonsaturation and nonequilibria

A central concept in the discussion of nonsaturation is that of a vacant or empty niche. Use of these terms has led to some controversy. It has been argued that a niche does not really exist unless it is used by an organism, and the term vacant niche is therefore meaningless. However, many studies have shown that some niches are much less utilized than other, similar ones. A vacant or empty niche, thus, is simply a concise way of saying that more species could exist in a habitat, as suggested by comparative studies. This is easiest to visualize by considering the spatial niche component, i.e., the habitat or microhabitat of species. If one freshwater pond is found to contain 10 species, and another of the same size and with the same characteristics 50, nothing prevents us from concluding that the former could accommodate more species, in other words that it contains vacant niches. However, how can we be sure that the characteristics determining species richness really are the same? In the case of ponds, it may be difficult, but such difficulties are much reduced or disappear altogether if we compare very simple habitats, for example fronds of fern or microhabitats on the gills or in the intestine of fishes. Selecting fish of equal size from identical habitats removes any error that may arise from overlooking minor differences that might be responsible for setting different "ceilings" to species diversity of parasites. The gills of some fish were found to harbour 1000s of parasites of close to 30 species, whereas most species of similar size and from similar habitats were found to have a few specimens of one or two species, or were not infected at all (for details see below). So, the evidence for the existence of vacant niches is convincing. There are of course reasons for the differences, but they are likely to be evolutionary, the result of historical contingencies. Given enough evolutionary time, more species may and will accumulate in the vacant sites.

Many studies have demonstrated nonsaturation of habitats, but nonsaturation does not necessarily imply nonequilibrium, the central theme of this book. If population size as a proportion of capacity is plotted against reproductive rates, the resulting bifurcation diagram never reaches the carrying capacity. In other words, equilibrium is established well below saturation (Figure 1.2), the consequence of a decreasing rate of population growth with increasing population size. The same can be expected for communities and ecosystems in evolutionary time: the number of species will always be below the maximum number possible (the "carrying capacity for species"), even if equilibrium conditions prevail at a particular point in evolutionary time. Therefore, in order to use nonsaturation as evidence for nonequilibrium, it must be shown that an increase in species numbers is likely. In other words, it must be shown that species numbers in ecosystems over evolutionary time have risen to the Recent (discussed in Chapter 6) and are likely to continue rising, and that species numbers can rise in communities as well, by comparing similar communities which differ only (or mainly) in species numbers.

Hutchinson (1957) was apparently the first to suggest that not all niches in a community may be filled, i.e., that vacant niches exist. Further references on vacant niches can be found in Rohde (1977b, 1979a, 1982, 1989, 1994a); Lawton (1982, 1984a,b); Price (1984); Compton *et al.* (1989); Begon *et al.* (1990); Lawton *et al.* (1993); Martins *et al.* (1995); Srivastava *et al.* (1997); and Cornell (1999). Walker and Valentine (1984) estimated that 12–54% of niches for marine invertebrates are vacant. Rohde (1998a) reviewed his earlier work on fish parasites: on 5666 fish of 112 species, a mean number of 4.3 species of metazoan ectoparasites on the heads and gills per fish species were found. The maximum number was 27 on *Acanthopagrus australis*, a small fish from warm-temperate waters in southeastern Australia, with most of the species strictly specific to this host species or a few related ones (Byrnes and Rohde 1992). The vast majority of fish species had fewer than 7, and 16 had none (Figure 2.6A). Assuming that 27 species is the maximum a host species can support (and there is no reason for this assumption: pathological effects are minimal) and that other fish species could support the same number (and there is no reason why they should not), only 15.9% of all niches are filled. Similarly, considering abundances (total number of all parasites of all species per number of fish of a particular species examined), maximum abundance was 3000 on *Lethrinus miniatus*, a medium-sized fish from the tropical Great Barrier Reef, but most fish had an abundance of fewer than five (Figure 2.6B). The mean abundance of 54.68 represented 1.82% of the maximum, again indicating that many more parasites could be accommodated. These data do suggest that niches are not only not saturated, but that they are very far from saturation. Great variability exists even within particular geographical zones (tropical, Antarctic, etc.), indicating that many (if not all) fish individuals could accommodate many more parasite species and individuals within each region as well. Morand *et al.* (1999) found a positive relationship between infracommunity species richness and total parasite species richness, and no evidence of ceilings, i.e., saturation, further support for the view that infracommunities of parasites are not saturated (see also Rohde 1994a).

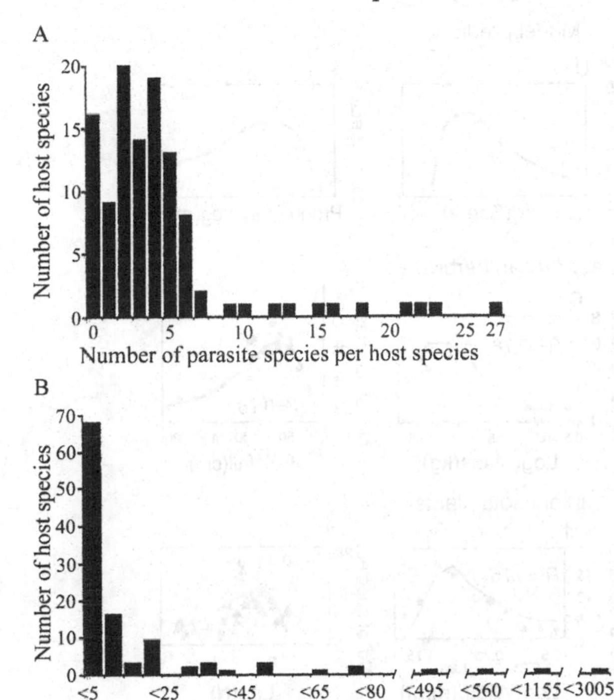

Figure 2.6 A. Number of species of metazoan ectoparasites on the heads and gills per species of marine teleost (5666 fish of 112 species). Note: maximum number 27 parasite species, most species with fewer than seven. If 27 is considered to be the maximum possible on all fish species, the percentage of empty niches would be 84.1%.
B. Abundance (= mean number of metazoan ectoparasites of all species per host species). Note: maximum abundance more than 3000, but most species with fewer than five. From Rohde (1998a). Reprinted by permission of the editor of Oikos.

Ritchie and Olff (1999) have proposed a new method to examine the problem of species packing. They used spatial scaling laws (fractal geometry) to derive a rule for the minimum similarity in the size of species that share resources, proceeding from the recognition that larger species detect only the larger patches of food but can tolerate lower resource concentrations than smaller species. The rule permits several predictions: (1) the body-size ratio of species of adjacent size should decline with increasing size of organism, because smaller patches richer in resources used by smaller species occupy relatively less volume than the larger but poorer patches used by larger species (Figure 2.7a); (2) there should be a unimodal distribution skewed to the left, when species numbers are plotted against the size of species, because smaller species have larger size

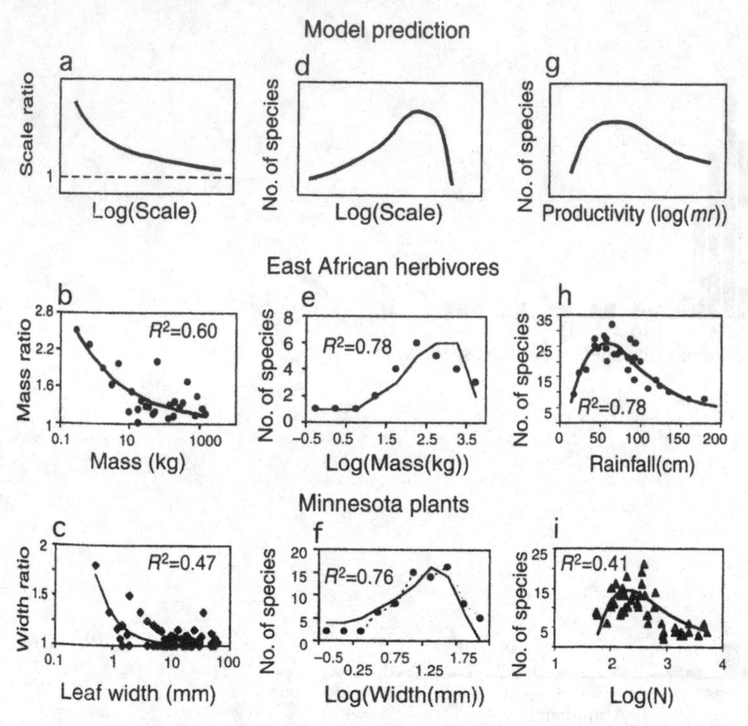

Figure 2.7. Ritchie and Olff have developed a scaling law model to test for species packing and have fitted data from some free-living animal and plant communities to it, concluding that these communities, are indeed densily packed and largely structured by interspecific competition. Predictions are shown in a, d and g: (a) Predicted size ratios (larger/smaller) for species of adjacent size versus size of the larger species. (d) Predicted number of species expected for communities versus log(size). (g) Predicted number of species versus productivity, or log(*mr*). Tests of the model are illustrated in b, c, e, f, h, i which show observed values for East African herbivores and Minnesota plants. From Ritchie and Olff (1999). Reprinted by permission of MacMillan Publishing Ltd., and the authors.

ratios and are therefore more loosely packed, whereas the largest species are limited by the maximum patch size (Figure 2.7d); there should be a unimodal distribution of species numbers versus productivity (Figure 2.7g). Serengeti (East African) mammalian herbivores and Minnesota savanna plants were shown to conform to the first two patterns (Figure 2.7b,c,e,f). The third pattern, according to the authors, is the most commonly observed pattern of species richness versus productivity, and was found for the two groups tested as well (Figure 2.7h,i). Rohde (2001a) has shown that the packing rules do not apply to metazoan ecto- and endoparasites of marine fishes. Parasites examined included Copepoda, Isopoda, Branchiura, Monogenea, Turbellaria, Trematoda, Cestoda, Hirudinea, Nematoda and Acanthocephala. Food (epidermis and blood for ectoparasites; epidermis, blood and gut contents for endoparasites) for these parasites is in unlimited supply and the only potentially limiting resource is space (Rohde 1991, 1994a). All species inhabiting the same habitat can therefore be considered as belonging to the same guild and were examined together. However, species of different large taxa (Platyhelminthes, Arthropoda) were also examined separately. Some fish species from the warm Pacific Ocean with large numbers of parasite species were selected, because of their likely effects due to dense packing. Also examined were parasites infecting a range of host species from various localities and latitudes jointly, including endoparasites of 1808 fish belonging to 47 species, justified since all fish species harbor similar parasites and a trend apparent in a component community of parasites (i.e., parasites of all species infecting one host population) should also be apparent or even augmented in a compound community (i.e., all parasite communities within an ecosystem). Both size of parasites, defined as the maximum length × width, and volume, (biomass) defined as maximum length × width × depth, were considered. In no case (with the exception of gill Arthropoda, which showed a very weak effect) did the packing rules of Ritchie and Olff apply. Examples are illustrated in Figure 2.8. The effect in the gill Arthropoda may be an artefact, due to the rarity of the very small males of some copepods which may have been overlooked.

These negative results support the view that parasites of marine fish do not live in saturated structured communities, but rather in assemblages not significantly structured by interspecific competition. Recent studies supporting this view are by Mouillot *et al.* (2003) who found a good fit of parasite data with Tokeshi's Random Assortment model, and by Poulin *et al.* (2003), who demonstrated a linear relationship between parasite biovolume and parasite diversity in parasite assemblages from 131 vertebrate host species, suggesting an additive effect and nonsaturation (Figure 2.9). The positive results presented by Ritchie and Olff (1999) for Serengeti grazing mammals and the North American savanna can be explained by the fact that they are either vagile (mammals) or disperse well (savanna plants), and that both utilize significant proportions of the resources for which different species compete, plants in the case of the former and light and space in the latter; interspecific competition is therefore expected (pp. 178–180).

Parasites, and this includes parasites of fish, represent a very large and probably the largest component of the Earth's fauna and should therefore

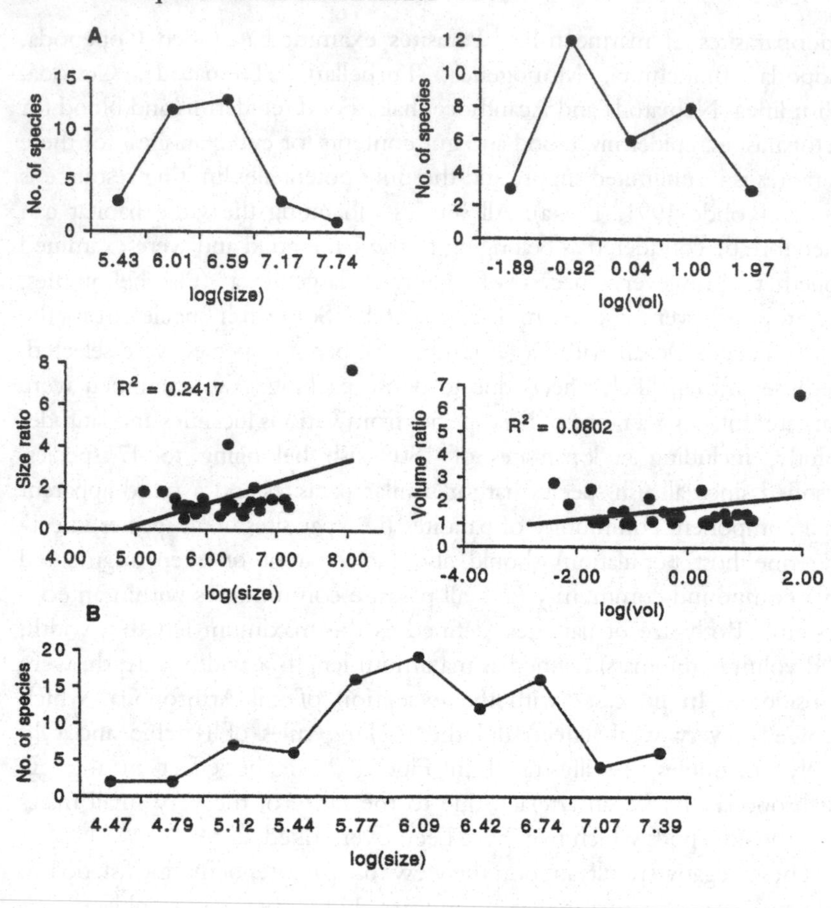

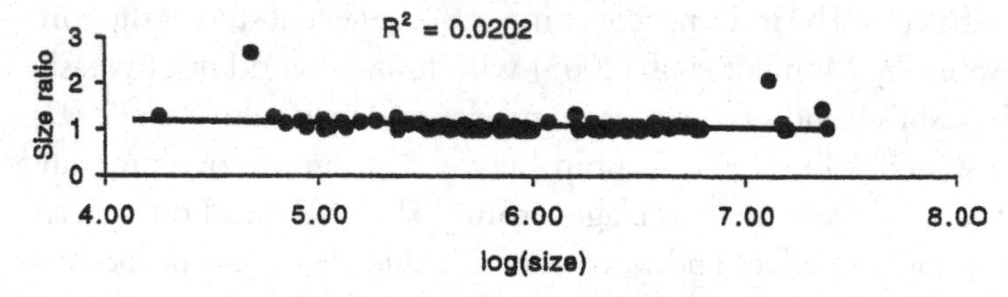

Figure 2.8. Tests of the scaling law model of Ritchie and Olff for species packing using ectoparasites of marine fish. (A) Metazoan ectoparasites of the marine teleost *Acanthopagrus australis.* (B) Metazoan endoparasites in the alimentary tract of marine teleosts (1808 fish of 47 species). It is clear that the scaling law model does not apply. Similar tests for a large number of parasite communities and from a variety of hosts likewise gave negative results. From Rohde (2001a). Reprinted by permission of the editor of Oikos.

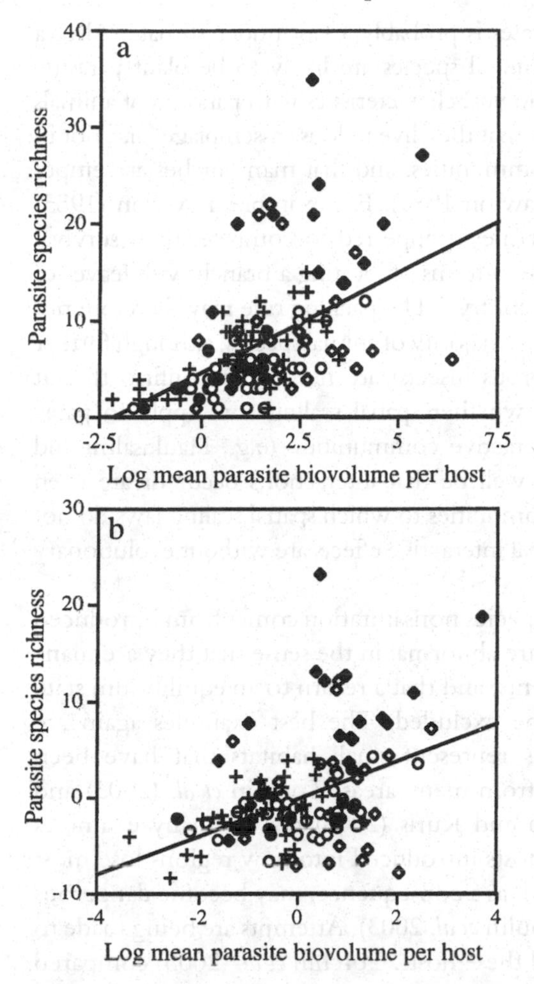

Figure 2.9. Relationship between parasite species richness and parasite biovolume across 131 species of vertebrate hosts. (a) raw data, (b) values corrected for sampling effort and host body mass. Crosses: fish, filled circles: amphibians, open circles: reptiles, open diamonds: mammals, filled diamonds: birds. The data suggest an additive effect and nonsaturation of habitats (hosts). From Poulin, Mouillot, and George-Nascimento (2003). By permission of the authors and Springer-Verlag.

not be ignored when determining the "mainstream" of ecological thought. There are approximately 21 700 species of fishes, 4000 of amphibians, 6550 of reptiles, 9200 of birds and 4237 of mammmals (various sources, Rahbeck 1993). The approximately 14 000 species of marine fish alone have probably more than 150 000 species of metazoan parasites, and a total of 300 000 parasite species for all vertebrates is probably a vast underestimate. Also, a great majority (millions) of animal species are likely to be plant-parasitic insects which share many ecological characteristics with parasites of animals (e.g., Rohde 1991). It is likely that they live in loose assemblages and not in densely packed interactive communities, and that many niches are empty (Lawton and Strong 1981, Lawton 1982). For example, Lowman (1985) concluded that canopy insects never appeared to compete, their survival rather "was a matter of chance in terms of locating a branch with leaves of palatable age, texture, and chemistry." The packing rule may therefore not apply to them, that is, to the vast majority of animal species, although further studies (especially of herbivorous insects) are needed to confirm this. It would also be useful to test whether spatial scaling laws apply to parasites living in supposedly interactive communities. (e.g., Haukisalmi and Henttonen 1993a,b). It may well be that interactions occur or are even common in nonsaturated communities to which spatial scaling laws do not apply, indicating that occasional interactive effects are without evolutionary consequence.

Convincing evidence for species nonsaturation comes from introduced species, although such cases are abnormal in the sense that they are man-induced and fairly recent events, and that a return to an equilibrium state in the near future cannot be excluded. The best examples again are parasites because their hosts represent small habitats that have been examined in large numbers from many areas. Torchin *et al.* (2003) and a recent review by Torchin and Kuris (2005) discuss many examples which show that, as a rule, hosts introduced into new regions lose most if not all of their parasites, and, as a consequence, may become dangerous pests (for an exception see Poulin *et al.* 2003). Attempts are being made to introduce parasites to control these hosts. Torchin *et al.* (2003) compared the parasite faunas of 26 host species of molluscs, crustaceans, fishes, birds, mammals, amphibians, and reptiles in their native and introduced ranges. 16 parasite species were recorded, on average, from native host populations. Only three of these, again on average, moved with their hosts into the new habitats, and a further four were newly acquired. A very well-examined marine example is that of the European green crab, *Carcinus maenas*, which has been introduced into several regions around the world, including the west coast of North America, and Australia. In its native habitats, it is infested by a variety of parasites; in particular by parasitic rhizocephalan barnacles (*Sacculina carcini*), which control the population size of the crab very effectively by castrating infected crabs. Rhizocephalans were not introduced into any of the newly colonized habitats. The reason is likely to be the means of transport: many if not most introduced marine animals are transported into new habitats in ballast water of ships, and larvae are not parasitised. Terrestrial examples include the cane toad introduced into Australia from Central America in an attempt to control native pests, but which has become a pest in its own right. It has none (or few) of the parasites infesting it in its native region, which may be a contributing factor to its success in Australia. Attempts are being made to control it by introducing parasites. There is no reason to exclude the possibility that the newly vacated niches on introduced hosts can be filled over evolutionary time; their present empty "status" is a historical contingency.

Finally, most assemblages can best be described by log series or lognormal distributions (e.g., Rosenzweig 1995). In other words, most species are rare and have little potential for interactions with other species. New (rare) species could therefore be added without much effect on species already present.

Kauffman (1993) presents a brilliant new look at evolutionary processes, which puts the existence of vacant niches in an evolutionary context and there now follows a brief outline of some of the main points of this approach. The approach emphasizes self-organization at the expense of natural selection, although the importance of the latter is also recognized. He says that "the order inherent in the busy complexity within the cell may be largely self-organized and spontaneous rather than the consequence of natural selection alone," and this also applies to other complex systems. Kauffman does not doubt the pre-eminent role of natural selection, but assumes that "spontaneous order is everywhere present" anyway. Evolution carries out adaptive walks towards peaks. "Morphology is a marriage of underlying laws of form and the agency of selection." He refers to the rational morphologists Goethe, Cuvier, and Geoffroy St. Hillaire, who searched for "some underlying logic or laws which would let us understand similar organisms as variations on some simple mechanisms that generate living forms." In Darwin's view, variations can occur in any direction and only selection decides which variation survives, although later authors have pointed out that evolution is constrained not only by selection. Kimura's (1983) strong formulation of "neutral theory" implies that all or most evolution at the molecular level is due to random drift among selectively neutral genetic variants. According to Kauffman, only two possibilities are open to current theory: selection and random drift. Concerning the effects of selection, it acts on systems that "spontaneously exhibit some particular form of order that is typical of an entire class of similar systems, called an ensemble";

"in sufficiently complex systems, selection cannot avoid the order exhibited by most members of the ensemble. Therefore, such order is present not because of selection but despite it"; and "many conceivable useful phenotypes do not exist." Phenotypic evolution in one species corresponds to a cloud of points (representing individuals) moving on a trajectory across phenotypic space. Assuming the null hypothesis that there are no constraints, a species can be thought as performing random-branching walks in a multidimensional phenotypic space, most of which is empty. Walks can be represented as occurring in rugged fitness landscapes. "In a fixed but rugged fitness landscape, radiation and ultimate stasis are utterly generic. We require no special mechanism to account for such phenomena" (such as initially empty niche space which is progressively filled). There is no need to assume filling of niche space or competitive exclusion to slow radiation (Kauffman, 1993, p. 77), because there are an ever decreasing number of directions close to adaptive peaks. Walks always stop at local optima below the global optimum.

In short, the existence of vacant niches is expected and widespread, and no special mechanisms are necessary to explain their existence.

We conclude not only that many (and perhaps all) niches are under-utilized by species, i.e., many vacant niches for further colonization exist, but also that nonsaturation is strong evidence for nonequilibrium in many communities.

3 · *Interspecific competition: definition and effects on species*

Interspecific competition is central to the equilibrium versus nonequilibrium debate. It is intuitively likely that under equilibrium conditions and in saturated niche space, competition will be more common and more intense than under nonequilibrium conditions and in nonsaturated niche space. In this chapter, I discuss evidence for and against the occurrence and significance of interspecific competition. The discussion begins with a definition of competition and its different types, and the main cause thought to be responsible for its occurrence, i.e., resource limitation. A discussion on the effects of competition on species in this chapter is followed by one on effects in communities in Chapter 4. The final section in Chapter 4 deals with general aspects and gives a conclusion.

Definition and types of competition, resource limitation as its main cause

Definition and types of competition

Interspecific competition results in the increase of the population density of one species at the expense of the reproductive rate and population density of another (Crawley 1986). However, it is not species but individuals within species that interact, and therefore, competition (at least as a starting point, but see p. 50) is better defined as an interaction between individuals that arises because of shared requirements for a limiting resource. It leads to reduced survival, growth and/or reproduction of at least some of the individuals (Begon *et al.* 1996).

Competition may be between individuals of the same species, i.e., it is intraspecific, or it may be between individuals of different species, i.e., interspecific. Competition may either be via exploitation of resources or by active interference (Park 1954). In exploitation competition, competing individuals use the same limited resource. In interference competition, some kind of activity may directly or indirectly reduce the access of a competitor to a resource. Nicholson (1954) distinguished contest and scramble competition. In contest competition, an animal is either fully successful or it is unsuccessful in utilizing a resource, i.e., the resource is completely utilized by one of the competitors. In scramble competition, the competitors attempt to utilize as much as possible of a resource to a certain minimum. In other words, competitors share a resource in various proportions (Barker 1983). Barker pointed out that both contest and scramble competition may lead to interference and provided further discussion of the relationship between contest and scramble, and interference and exploitation competition.

The foregoing discussion has shown that in cases where competition occurs, it should be for some resource, whether it be space, food, or some other resource that is in limited supply. The role of resource limitation is discussed in the following, and in particular the question of how solid is the evidence for its importance.

Resource limitation

Supposed competition for resources is widely used to explain ecological patterns. However, Levin (1970) has shown that resource limitation of all species in a system is not necessary for extinction to occur. Specifically, even two species that feed on different food resources that are not in limited supply cannot indefinitely coexist if they are limited by the same predator. This is because each species increases when the predator decreases, and vice versa, and since each species must have a different threshold predator level at which it stabilizes, the species with the higher threshold will replace the other. Therefore, the only important criterion for species coexistence is that limiting factors (whether food resource, predation, etc.) differ and are independent.

The Red Queen hypothesis (Van Valen 1973) (the same as the Rat Race hypothesis, Rosenzweig 1973) assumes that resources are limiting and that the most important component of a species' environment is the species with which it interacts. Therefore, changes in any of the interacting species modifies the environment of all the other species with which it interacts, and even if there are no changes in the abiotic environment, species will still evolve.

Various models that have been established to fit species abundance patterns are based on the assumption that species' abundances are determined by resource allocations (e.g., Tokeshi 1990, 1999). One of these models, the Random Assortment model, was fitted successfully to data for three parasite communities of fish from the Chilean coast. This means that interactions between species are not important, and that these parasites probably live in unsaturated communities (Mouillot *et al.* 2003). The packing rules based on fractal geometry are also based on the assumption that species compete for resources in limited supply. They apply to herbivorous mammals competing for food and to savanna plants competing for light and space, but they do not apply to all parasites (see pp. 41–46, 76).

Lawton (1984a), in his study of community organization, refers to a number of studies that suggest that certain ecological patterns may be entirely or partly determined by interspecific competition, but he points out that each of these patterns can also be explained by other hypotheses. A search for competition should follow demonstration of interspecific resource limitation, but such limitation has, for example, not been found in insects inhabiting bracken, which are rare and, therefore, unlikely to compete for resources. But Lawton admits that this does not exclude evolution to minimize interspecific competition in the past. Wiens (1984) points out, with regard to "ghost of competition past," that such hypotheses are not testable, and that demonstration of patterns of resource partitioning does not say anything about the processes leading to such partitioning (see also Connell 1980 and pp. 85–89). Both interactive and non-interactive communities may contain species that partition resources. He cites examples showing greatest resource overlap at times of scarcity, others at times of abundance (p. 423). "The introduction of even modest levels of spatial heterogeneity into resource systems potentially permits an almost unlimited number of species to coexist at equilibrium on a restricted set of resources"; "preconceptions about the systems must be recognized as such, and not permitted to bias the research design or interpretation of results."

Price (1984) emphasizes that the resource base of a community has to be clarified, before community phenomena can be examined. He categorizes resource types as follows: rapidly increasing, pulsing or ephemeral, steadily renewed, constant, or rapidly decreasing, and gives examples: e.g., endoparasites use steadily renewed resources, constant resources are physical ones, such as surfaces on which intertidal organisms can settle, etc. Population responses to resource availability depend on generation times, and permit prediction on whether resources will be in short supply and whether competition will occur. Spatial distribution of resources (whether uniform or patchy) will also permit predictions. However, since most resources are patchy, and because patchy resources have a low probability of colonization, Price assumes that interspecific competition is not likely. Furthermore, disturbances will disrupt interactions between species, thus resulting in nonequilibrium conditions. Interspecific competition is most likely when at least two potential competitors are likely to co-occur in the same patch, and when population responses are rapid leading to exhaustion of many of the resources. Price (1984) predicts that intestinal helminths of vertebrates should show little competition, as resources are constant (space) or steadily renewed (food), hosts are patchy and population responses of helminths slow (no reproduction in the host).

A review of the findings on fish parasites as a general model for evaluating intra- and interspecific interactions in low density populations in resource-rich habitats was given by Rohde (1991). He used ectoparasites of marine fish as a model but extended the conclusions to other taxa, concluding that most animal species are likely to live in low-density populations in resource-rich habitats. In such habitats, many potential niches are vacant and interspecific competition is of little importance although it may occur. Microhabitats may or may not expand when population density is high, and niche restriction may have the function to enhance chances of mating. Segregation often may not be due to competition but reinforcement of reproductive barriers (see pp. 85–89 for further details).

Nevertheless, there can be no doubt that resources are sometimes limited and that species compete for them. Brown *et al.* (1979, further references therein) have shown experimentally that competition for seeds is important in determining community structure of desert granivorous rodents, and Pimm (1978), also experimentally, has shown that, in hummingbirds, competition occurs when resources are predictable, but decreasingly so when resources became less predictable.

The finding that resources are sometimes and perhaps often limiting, leads to the question of how similar species competing for the same resource can be, a problem discussed in the following section.

Effects of competition on species

Limiting similarity and ecological character displacement

If resources are indeed limiting, competition for similar resources should lead to some displacement between species in order to reduce competition. But how close can two competing species be? Hutchinson (1959) believed that there was a body size difference (in length units) of 1:1.3 in coexisting species pairs, indicative of the difference necessary for species to coexist at the same level of the food web. However, is divergence between species really necessary to permit coexistence? As pointed out by Rosenzweig (1995), the original deterministic proposal by MacArthur and Levins (1967), that two species, in order to survive, must differ to a certain degree, is wrong. As already realized by May and MacArthur (1972), the limiting similarity in such a deterministic system is zero, i.e., even very similar competitors will not become extinct. Hubbell and Foster (1986, cited by Rosenzweig 1995) showed that extinction, even in identical species, may take a very long time, and in populations of a few thousand, extinction time may be as long as speciation time. Hence, Gause's principle is wrong. This is not so in stochastic systems: extinction may still take a long time, but a species too similar to the competing one will finally be pushed over the rim by some "accident" (in Rosenzweig's words, see also Crawley 1986). Also, the values of limiting similarity change according to environmental conditions and properties of the niche (Abrams 1983). Nevertheless, because systems are not strictly deterministic, because they are stochastic to a certain degree, either extinction or divergence due to competition is at least a possibility.

An early example of an explanation of differences between species as the result of "ecological character displacement" (Grant 1975) is provided by Grinnell (1904), who postulated that more than one species can coexist in the same habitat only by using different sorts of food and using different ways of getting the food. Brown (1975), in his paper on the geographical ecology of desert rodents, concludes that "competition is a major force in the structuring of rodent communities. Coexisting species subdivide food resources by collecting seeds of different sizes and by foraging in different microhabitats." Character displacement in body size is brought about by selection of different seed sizes. When competitors are absent, species use a much wider range of seed sizes than if they are present, and species that have invaded isolated habitats occur in exceptionally high population densities. One of the best examined examples of ecological character displacement is that of Darwin's finches on the Galapagos Islands. Grant and Schluter (1984) examined beak morphology of various species and demonstrated that species co-occurring on the same island differ more strongly in beak morphology than the same species occurring singly on different islands. They concluded that interspecific competition was responsible for the greater differences in sympatric species. Size differences of the species was considered to correspond to size differences in food particles. Studies of birds in New Guinea and elsewhere were also used to support this hypothesis (Diamond 1973, 1975; Cody 1974; discussion in Roughgarden 1989). An example of supposed ecological character

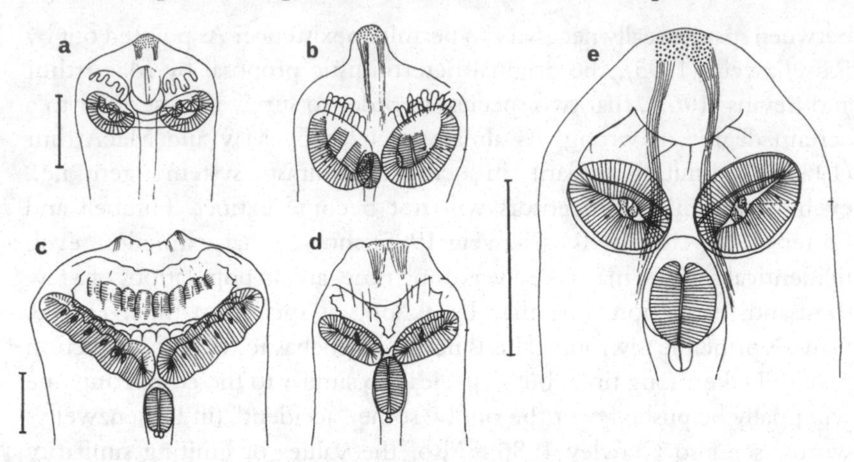

Figure 3.1. Anterior ends of five species of polyopisthocotylean Monogenea from the gills of *Carangoides emburyi* (a,b) and *Scomberomorus commerson* (c to e) on the Great Barrier Reef, Australia. All species are blood feeders and live in identical or largely overlapping microhabitats on the gills of their respective fish hosts, i.e., they exploit the same resource. Nevertheless, there are marked differences in the sizes of pharynx and oral suckers. Scales 0.1mm. From Rohde (1979b). Reprinted by permission of Blackwell Science Ltd.

displacement in parasites was discussed by Butterworth and Holmes (1984): two species of trematodes of the racoon, *Pharyngostomoides procyonis* and *P. adenocephala*, show character displacement (size differences) when they co-occur in the same geographic area, interpreted as meaning that they use different food particles.

However, importantly, character displacement can have causes other than competition. A particularly important aspect of character displacement is the difference in size of feeding organs in sympatric species, supposedly to avoid competition for the same resource (see discussions in Christiansen and Fenchel 1977; Schoener 1986b; Eadie 1987). However, which null model does apply? Is it possible that differences are fortuitous? Andrewartha and Birch (1984) cite a paper by Pulliam and Enders (1971) who found three to five species of finches in the same habitat in the southeastern United States in the summer. The species differed in the size of their beaks but there was almost complete overlap in the size of seeds eaten. Also, seed-eating heteromyid rodents seem to eat seeds of identical size, and seed husking efficiency seemed identical (Rosenzweig and Sterner 1970). Rohde (1979b, 1991) has demonstrated that the feeding organs (pharynx, oral suckers) of monogeneans infecting the gills of the same host species and using the same food, blood, differ in size and shape (Figure 3.1). Hence, he concluded the differences may be fortuitous, which does not exclude the possibility that they may have an adaptive value, i.e., enabling the species to extract the blood from different parts of the gills, although this is not necessarily due to competition for feeding sites. Most importantly, character displacement may be the result of reinforcement of reproductive barriers. This is discussed fully in Chapter 5.

A more general objection to the approach explaining differences between species by competition for limited resources resulting in character displacement was made by Andrewartha and Birch (1984), who point out that one should expect to find similar species in similar habitats. One should rather ask: why can species with different ecological requirements live together? Another important point to be established is how solid the evidence given for ecological displacement really is, a question discussed next.

Habitat segregation as evidence for interspecific competition

Niches of sympatric species are never strictly separated, as shown by two of many examples: five abundant tropical terns breeding sympatrically showed some degree of segregation in their diets and foraging ranges, but there was nevertheless much overlap (Surman and Wooller 2003), and two sympatric jackals in northwest Zimbabwe do not have clear-cut niches and coexistence is probably possible because of extreme flexibility in diet and behaviour (Loveridge and Macdonald 2003). Overall, it is highly unlikely and probably impossible that two species overlap exactly in their niches including habitat, and a null model should be based on the assumption that such differences are fortuitous. Nevertheless, many authors have explained habitat segregation of species by interspecific competition, without giving evidence or considering other explanations. Rohde (1989) drew attention to alternative explanations for habitat segregation in helminths, usually explained by interspecific competition. For example, Schad (1962, 1963) examined the distribution of eight species of the nematode genus *Tachygonetria* (three subsequently transferred to *Mehdiella*) in the intestine of the Greek tortoise, *Testudo graeca.* Four species had a paramucosal distribution, the other four were found in the lumen; species in each group were segregated to various degrees although all showed some overlap. Species differ in body size, length of the male copulatory sclerites and male genital papillae, etc. If species occurring in the lumen are arranged in order of overlap, the series *Tachygonetria numidica, T. macrolaimus, T. dentata,*

T. conica is obtained. These species differ in spicule length as follows: approximately 55, 120, 40, 120 μm. It seems that species with the greatest spatial overlap are reproductively segregated by copulatory spicules of different size. The situation in paramucosal species is not as clear: *M. uncinata*, *M. microstoma*, *T. robusta*, *M. stylosa*, have spicules of 100, 100, 100, 550 μm length. However, in addition to spicule length there are differences in body length and in genital papillae around the male gonopore which may lead to reproductive isolation; and the possibility must also be considered that unstudied chemical factors may contribute. Evidence, although not conclusive, shows that the interpretation by Schad (1963) of segregation as the result of competition is at least doubtful.

Species of *Drosophila* and related genera have been used in many field and laboratory experiments, some concerned with the question of niche segregation and its causes. Barker (1983) has reviewed some of these studies and concluded that, although the idea of interspecific competition leading to niche segregation and differential adaptation is attractive, and although studies have demonstrated niche segregation, they provide no evidence for the mechanism that has led to segregation.

Importantly, random selection of microhabitats in largely empty niche space may also lead to niche segregation, even if interspecific effects have never occurred in evolutionary time and are not occurring now (Rohde 1977b).

If competition has the effect of limiting the similarity between species living in the same habitat, it might lead to a reduction in the co-occurrence of congeneric species in sympatric habitats. Is there evidence for this assumption?

Reduction in the number of sympatric congeners as evidence for competition

A reduction in the number of sympatric congeners relative to the number expected, if congeners were randomly acquired, has been suggested as an indicator of the evolutionary significance of interspecific competition (Pianka 1983). Elton (1946) appears to have been the first to make this suggestion. He counted the numbers of congeneric species in many well defined habitats and compared them with the total number of congeners in the various genera. He found a much smaller percentage of such species in the various habitats than in the genera, and concluded that interspecific competition was responsible for the discrepancy. Concerning parasites, related species supposedly compete more strongly than unrelated ones

(Goater *et al.* 1987), because their ecological requirements are closer and they therefore have a greater potential for resource competition, which should lead to a reduction of sympatric congeners.

But what is the null model? How many congeners are expected if they are acquired at random? Congeners are extremely common in some habitats, but not in others. Particularly striking are the large numbers of congeneric parasite species infecting the same host, for example flocks of nematodes in Australian marsupials. Thus, up to 20 species of the genus *Cloacina* (Nematoda, Strongyloidea) occur in a single species of kangaroo (Beveridge *et al.* 2002). Parsimony analysis of the nematodes showed that the species flock in each host species is polyphyletic in origin, although a few related parasite species occur in some of the hosts. Interestingly, even closely related host species may differ considerably in numbers of congeneric nematodes. For example, *Macropus agilis* is most commonly infected with a single species of *Cloacina*, whereas *M. dorsalis* normally harbours a large number. But does the small number of congeners in the former species really indicate competitive exclusion? If this were the case, why are there so many species in a closely related host? There is certainly no evidence whatsoever that interspecific competition in one host should be stronger than in another. The most parsimonious explanation for the differences appears to be that they are fortuitous, due to some ecological differences between host species that make infection more difficult in one than in the other. For negative results to find reduction of sympatric congeners, see also Pianka (1973) for lizards and Terborgh and Weske (1969) for birds.

Concerning fish parasites, Rohde (1989) pointed out that a null-hypothesis, i.e., how many congeners could be expected, cannot be established, but the great number of congeners found makes it unlikely that a reduction in their number due to competition has occurred. Data are incomplete, but it seems likely that fish species without congeneric parasites are the exception rather than the rule.

Nevertheless, in some cases a reduction in the number of congeners co-occurring in the same microhabitat has been demonstrated. Rohde and Hobbs (1986) have shown that congeneric monogeneans on the gills of marine teleost fish are more strictly segregated than non-congeners. However, such segregation of congeners occurred only in species with identical copulatory organs. Congeners with distinctly different copulatory sclerites often share the same microhabitat. This means that not interspecific competition, but reinforcement of reproductive barriers is responsible for segregation. This is discussed in detail in Chapter 5.

Effects of competition on microhabitat width

Microhabitat width is just one of the many niche dimensions that may be affected by competition. Effects may occur in extant communities, e.g., when the microhabitat of one species contracts in the presence of another, or it may have occurred in the past as the result of coevolution with competitors (the "ghost of competition past").

Rohde (1979a) gave a detailed discussion of evidence on niche restriction in parasites and concluded, with respect to microhabitat restriction, that

(1) interspecific competition in parasites occurs and may lead to competitive exclusion or changes in microhabitat width in some or all co-occurring species (interactive site segregation), but there is no evidence that such effects lead to evolutionary changes and avoidance of competition, i.e., to selective site segregation;
(2) parasites with coinciding or overlapping microhabitats often show no interactions;
(3) related species commonly have widely overlapping microhabitats;
(4) the effect of species intrinsic factors (facilitation of mating) on niche restriction is indicated by the finding that competing species often do not exist and cannot have existed in the past;
(5) circumstantial evidence suggests that niche restriction may be due to selection for increasing intraspecific contact and thus mating ("mating hypothesis of niche restriction");
(6) the probability that two species have completely coinciding niches is infinitesimally small, and niche differences should not be used as evidence for competition.

A more detailed discussion of the evidence is given on pp. 81–85.

As pointed out above, one supposed effect of interspecific competition is change in microhabitat width in the presence of competing species. Kennedy (1985, 1992), in careful studies, demonstrated that microhabitats of acanthocephalans in the digestive tract of the eel *Anguilla anguilla* in Britain contracted in the presence of other species. In contrast, Rohde (1981a) compared microhabitat width (as measured by the number of sections of the gills and mouth cavities of approximately equal size) of ectoparasites of poorly and heavily infected marine fish. He found no correlation. Moreover, there is no significant difference in the number of host species used by monogenean ectoparasites of marine fishes at different latitudes, although tropical fish have considerably greater numbers of

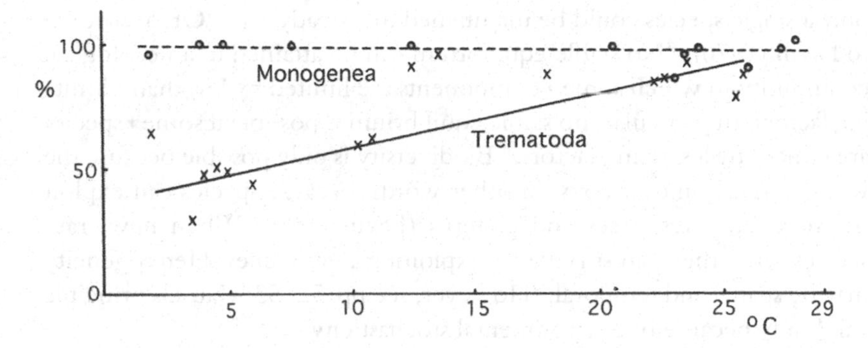

Figure 3.2. A comparison of host ranges of ectoparasitic Monogenea and endoparasitic Digenea (trematodes) of marine teleost fish at different latitudes. Host ranges of gill Monogenea (o) and trematodes (x). Ordinate: % of species in a survey restricted to one or two host species. Abscissa: mean annual surface temperatures in °C. Host ranges of Monogenea are more or less the same at all latitudes, whereas host ranges of trematodes are much wider at high than at low latitudes. However, the apparent increase in host ranges at high latitudes in trematodes disappears when correction is made for intensity and abundance of infection in different host species, i.e., parasite species have similar host specificity at all latitudes. From Rohde (1978c). Reprinted by permission of Springer-Verlag.

species than cold-water fish (Figure 3.2). Sample size was not responsible for these findings. Prevalence of infection tends to be greater in tropical fish. In other words, niche width (at least along the dimensions, host range, microhabitat width, and prevalence of infection) is not narrower in species-rich communities. One reason is that many niches (microhabitats) are empty. Also, many and perhaps all host species could accommodate more parasite species, and fish with few ectoparasites do not have more endoparasites, which makes the argument invalid that a host is a functional unit and that the carrying capacity of a host with few ectoparasite species might be used by parasites in other habitats.

Competitive exclusion and its prevention by spatial/temporal heterogeneity

A direct consequence of limiting similarity is competitive exclusion, i.e., the complete displacement of one species by another, competing species. This is expressed as Gause's principle, according to which (in its original and widely used form) species using the same resource cannot coexist (see previous sections), (e.g., Levin 1970; May 1981). In accordance with this principle, in an absolutely homogeneous world with a single resource, only a single species could be maintained in a steady state. Or, according to Levin (1970), "No stable equilibrium can be attained in an ecological community in which some *r* components are limited by less than *r* limiting factors. In particular, no stable equilibrium is possible if some *r* species are limited by less than *r* factors." Biodiversity is only possible because the world is not homogeneous, in other words, because species can exploit heterogeneous resources and patterns (Levin 2000). When new, rare species arise, they can survive by exploiting new niches. Heterogeneity may be spatial and temporal. (However, see pp. 52–53: Gause's principle valid only because of environmental stochasticity).

Leslie *et al.* (1968) demonstrated competitive exclusion in two *Tribolium* species. However, they also noted unexplained coexistence between the species. As an explanation of this observation, Edmunds *et al.* (2003) have recently proposed a model in which, as interspecific competition increases, there is a sequence of bifurcations; that is, a scenario with two stable competitive exclusion equilibria is replaced by a scenario with two competitive exclusion equilibria and a stable coexistence cycle. In other words, two species may well coexist on one limiting resource, even if (or because) there is increased competition.

An example of competitive exclusion is the trematode *Gorgodera euzeti* and the monogenean *Polystoma integerrimum*, both infecting the urinary bladder of the frog *Rana temporaria* in the Pyrenees (Combes 2001, reference therein). The number of frogs examined was 1941, the number infected with the first species alone was 576, with the second species 280, with both species 39. If double infections had occurred by chance alone, the number should be at least $576 \times 280/1941 = 83$. This example seems convincing, but can the possibility be excluded entirely that slight differences in habitat preferences of the two species are responsible for the smaller than expected number of double infections? Combes (p. 434) gives some other examples for parasites (see also Kuris 1990, for larval trematodes). For a detailed discussion of competition in larval trematodes in snails, demonstrated by experiments or inference from field studies, see Combes (2001), page 428 onwards (for details see Chapter 8). Rohde (1979a) pointed out that intraspecific crowding may have effects similar to competitive exclusion. Moulton and Pimm (1987) studied the effects of the 49 bird species introduced to the Hawaiian Islands from 1869 to 1983 and found convincing evidence for competitive exclusion.

Competitive exclusion in a particular habitat can be overcome by immigrations from adjacent habitats. Already Gause (1935), in his classical study on competition between *Paramecium caudatum* and *P. aurelia* and predator–prey dynamics of *Paramecium* and *Didinium*, pointed out that coexistence of both species of *Paramecium* was made possible by periodic immigrations from adjacent habitats (not possible in simple laboratory experiments which did not replicate the spatial component of natural habitats). Huffaker (1958) showed this experimentally in a phytophagous mite–predator system. Tilman *et al.* (1997) and Hanski (1997) discuss several models that simulate population dynamics in spatial habitats (see also other chapters in Tilman and Kareiva 1997).

Ecosystems are complex adaptive systems (CAPs), and resilience in such systems results from maintenance of heterogeneity (Levin 1998, references therein). It is important to note that not only spatial, but also temporal heterogeneity, contribute to diversity and resilience of systems. It is also important to note that resource heterogeneity may exist in "essential homogeneous" environments, i.e., in environments that, in the long term, have similar statistical characteristics, but differ due to historical events such as local disturbances caused, for example, by the extinction of a dominant species (Levin 1998).

In simple laboratory experiments, the number of potentially competing species has to be small. However, in nature, the vast majority of ecological systems comprise a multitude of species, and effects of competition are not necessarily the same in species-poor and species-rich systems. The next section deals with the question of how diversity affects species interactions. It also discusses the effects of supposedly "harsh" and "benign" environments, and various other factors that may have an effect on the degree of competition.

Factors that determine the degree of competition in populations

(1) Species diversity and competition, and the effects of environmental factors

The probability of coevolution of a species with competitors may be inversely related to species diversity. For example, corals and trees in tropical rainforests have the potential to interact strongly not with one but with many neighbors (Connell 1980). In Queensland rainforests, the mean number of tree species with a height greater than 0.5 m per 10 × 10 m plot had 57 other trees among the nearest neighbors, and for some of the less common species, all nearest neighbors were different species. Coevolution between them is therefore unlikely.

Connell (1975) points out that interspecific competition is most likely in moderately harsh physical environments. Very harsh conditions reduce population densities to such a degree that competition becomes unlikely, and in benign environments parasites, predators, and herbivores become more effective, also reducing population densities. As a consequence, coevolution between competitors is most likely in moderately harsh environments.

(2) Nonequilibrium reduces importance of competition

In a wide variety of community types (avian, insect, parasite, fish) non-equilibrium conditions were shown to reduce the effects of interspecific competition, or competition may not exist at all (e.g., Dayton 1971; Wiens 1974; Sale 1977; Connell 1979; Grime 1979; Price 1980). This phenomenon is widespread and well supported by observations. Examples are discussed in various chapters (e.g., pp. 34; 37; 105–108).

(3) Population density and competition

Andrewartha and Birch (1954) point out that the terms "competition" and "density-dependent factors" are often used as synonyms. They quote Nicholson (1933), and Varley (1947) who writes: "The controlling factors which keep a population in balance must be affected in their severity of action by the population density on which they act." Note the "must". Andrewartha and Birch, in their detailed account and dissecting many examples, conclude that the exaggerated emphasis on competition is fallacious. Most natural populations never become sufficiently numerous to use a substantial proportion of the resources needed by them. Density-dependent factors and competition therefore do not become operative. They list three possibilities of how numbers of animals in natural populations can become limited. They are: (1) shortage of resources; (2) inaccessibility of resources; and (3) shortage of time during which an increase in reproductive rates is positive. Based on their very great experience with various animal populations, they conclude that the first is probably least and the last most important in nature. Importantly, the authors point out that populations which are usually studied are atypical, i.e., they are common and not rare like the vast majority. An excellent example for this point are the studies of larval trematodes in snails. The impression is given that most snail populations are heavily infected with many parasite species. However, many (and probably most) marine snail species have no or few parasites, and the same refers to terrestrial and freshwater snails. Authors don't bother with uninteresting, poorly infected host populations, they rather concentrate on the more interesting, richly infected populations (see Chapter 8).

(4) Competition is stronger between core than satellite species
A special case of the effects of population density on the intensity of competition is that of core versus satellite species. Core species are more prevalent and found in high population densities, satellite species are much rarer. Holmes (1987) points out that competition becomes more likely with increasing probability of two species encountering each other, i.e., it is more likely in core than in satellite species. This intuitively makes sense but these two types of species are not always distinguishable.

(5) Heterogeneity of populations and competition
Andrewartha and Birch (1984) give a criticism of competition theory, which states that authors providing evidence for the effects of competition usually assume that a population is a homogeneous unit, which is totally unrealistic when considering natural populations. But even in local populations, evidence does not support the view that competition leads to extinction.

Does competition play a role in the formation of species? Such a role was, for example, postulated by Rosenzweig (1995). But how solid is the evidence?

Competitive speciation

Allopatric speciation is generally accepted as an important, and perhaps the most important, kind of speciation. But there is increasing circumstantial evidence that sympatric speciation is important as well (see recent review by Via 2001). Sympatric speciation may occur by polyploidization, for which there are many examples in plants, but there also may be non-polyploid sympatric speciation (named "competitive speciation" by Rosenzweig 1995, who defines it as expansion from a single ecological opportunity into a new opportunity that lacks competition). Rosenzweig discusses many examples and gives references. See for example Tauber and Tauber (1977, 1987) for insects (also Hendrickson 1978; Tauber 1978). Rosenzweig evaluates the data and concludes that all three modes of speciation (by polypoidy, geographical speciation and competitive speciation) are significant.

Evidence for competitive speciation is similar to that often given for habitat shift due to competition, ecological character displacement, etc., i.e., that the new species uses different resources. For all these phenomena different explanations are possible, and this includes at least some of the cases given as evidence for sympatric competitive speciation. If one or a few mutations lead to the establishment of a species co-occurring with the parent species, it must be reproductively isolated from the parent in order to survive as a species. Such isolation may be spatial, or it may be by different copulatory organs, different pheromones, etc. Whatever the case, differences in the use of resources, if they exist, may be purely secondary. In other words, in such cases not competition but reproductive segregation may be the driving force for speciation. Competitive speciation can be assumed only if explicit evidence for competition as the causative factor is given. But how can we test for such evidence?

Tests for divergence due to competition

Dissimilarity between species is often used as evidence for competition (see pp. 52–55), but Connell (1980) has convincingly shown that tests for competition are inadequate. In spite of his criticisms, many authors still uncritically claim competitive effects without any evidence whatsoever. Connell established very strict criteria for demonstrating divergence of competitors: (1) there actually has been divergence in resource use between competitors; (2) competition and not some other mechanism was reponsible for the divergence; and (3) divergence has a genetic basis and is not simply phenotypic.

Concerning the first point, Connell stresses that divergence between extant species has not been demonstrated except for some pests of crop plants; some fossil sequences showing divergence are known but it is impossible to say whether competition was responsible. Concerning the other two points, field experiments are necessary to provide evidence, but these have been performed only in part, for one case. Connell concludes that, at present, there is little support for the coevolutionary divergence of competitors, and that is probable only in low diversity communities. According to Connell, point 2 can best be tested by manipulating the distribution and abundance of one or both species of competitors in field experiments. However, controls are difficult. Thus, if all populations that are compared are sympatric, it is impossible to obtain data on the relevant dimensions in multidimensional niche space, and limiting similarity is purely theoretical and has no proven biological basis. Also, even if species overlap in resource use, they may not compete for them. On the other hand, if there are allopatric and sympatric populations and if they do indeed differ, it is difficult to prove that the only difference is the absence of the competing species in allopatric situations, and comparison with random assemblages is very difficult.

Further, according to Connell, a direct measurement of competitive effects would be transplantation of individuals from an allopatric population into a sympatric location, and observation of the effects where a competitor is present and where it has been removed. Individuals that are not transplanted but handled and observed in the same way as the transplanted ones could serve as controls. Competitive effects would be shown by compression of the niche in the presence of competitors. Finally, a genetic basis for the differences would have to be demonstrated. These criteria are very strict and very difficult to apply. Therefore, many authors have used plausibility arguments. Colwell (1984) and Cooper (1993) emphasized the usefulness of plausibility arguments based on background information. However, it is important not to forget that such arguments may easily mislead if not applied very critically. If not used critically they may, for instance, lead to a confusion of interspecific competition with other factors such as reinforcement of reproductive barriers responsible for species segregation.

Nonlinear dynamics (chaos) in populations and the outcome of competition

A considerable amount of work has been done on the non-linear dynamics of populations in one- and multi-species systems. Here, we discuss some selected models in order to show that the outcome of competition between species may be largely unpredictable.

Hawkins (1993) used a simple model incorporating local dynamics and dispersal for studying interactions between dispersal, dynamics, and chaos. His aims were to show that the scale at which density dependence can be detected, depends strongly on the intensity of dispersal between patches, that dispersal plays a stabilizing role, and that there may be extreme, sensitive dependence of long-term behavior on initial conditions.

Hawkins refers to Park's experiments on *Tribolium*, which had been interpreted as showing that the outcome of competition was due to initial conditions, or – in cases where it was not – the result of stochastic processes. However, the model established by Hawkins showed that a simple deterministic process, and not stochasticity, may be the explanation for such cases. This is apparent even when numbers within each patch are very small. Hawkins concludes that "This extreme dependence of qualitative behavior on historical effects calls into question many of the paradigms underlying ecological explanations." Hawkins' results further show that the appropriate spatial scale must be selected to demonstrate density dependence. If the coupling between two patches is weak, density dependence can only be detected for a single patch, but not the whole population. The dynamics of the whole population, under such conditions, is similar to the "density-vague" dynamics discussed by Strong (1984). Hawkins (1993) quotes several authors who have indeed shown that density dependence in a number of insect populations can be detected only at very small spatial scales. Overall, dispersal has an important function in stabilizing population dynamics. Rohde and Rohde (2001) used a different ("fuzzy chaos") model in which metapopulations were assumed to consist of subpopulations with different reproductive rates. They demonstrated that chaos, apparent in subpopulations, is largely reduced in the metapopulation. If applied to two- or three-species systems, the model shows that the outcome of competition between two or three species is to a large degree unpredictable (for two species see Figure 3.3, for 3 species Rohde 2005b). A similar conclusion was reached by Neubert (1997) for single-species systems. His model exhibits a "riddled basin of attraction," i.e., every point in one basin is arbitrarily close to another basin. Or, in other words, every initial condition is on the boundary between different basins of attraction, which makes predictions impossible. As Hawkins (1993) earlier, Neubert refutes the explanation of Park's *Tribolium* experiments as the outcome of stochastic events. The fact that sometimes one species and sometimes another wins may not be due to stochasticity, but are examples of "simple deterministic systems that have .... stochastic outcomes."

Null models and the demonstration of interspecific competition

Recently, powerful mathematical tools (null models) have been applied to investigating the role competition plays in structuring communities. A null model used to show whether type II communities are indeed the result of interspecific competition is discussed in the section on pp. 73–75, and a null model analysis of co-occurrence patterns of ectoparasites of marine fish is discussed on pp. 125–127. A comprehensive and thorough treatment of null models in ecology is provided by Gotelli and Graves (1996), and Connor and Simberloff (1986) have given a particularly lucid and thoughtful discussion of the significance of competition, scientific method, and null methods in ecology. They draw attention to the difficulties in providing secure evidence for competition, because experiments are difficult to perform or even impossible, and nonexperimental evidence may be difficult to interpret. They discuss

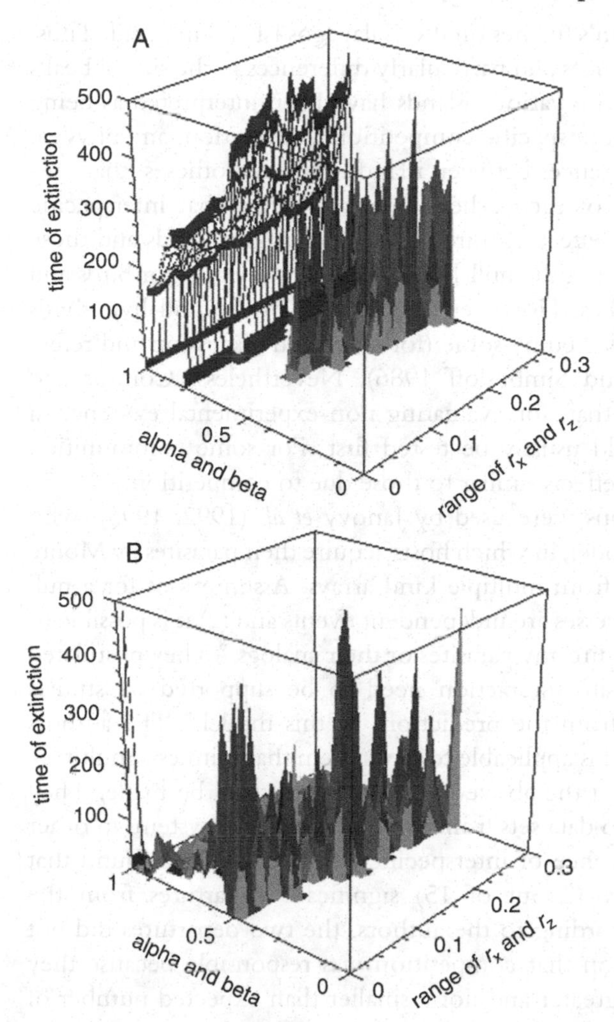

Figure 3.3. Fuzzy chaos modelling is based on the assumption that populations consist of subpopulations that are largely separate and differ in their reproductive rates. In this example, we apply fuzzy chaos modelling to two interacting species. Each species is assumed to consist of 10 subpopulations with different reproductive rates. The ranges of the reproductive rates of species x and z, r_x and r_z (indicative of the heterogeneity of the metapopulations) are plotted against alpha and beta (the competition factors of both species), and the time of extinction in number of generations. The initial population sizes of species x and z, x_0 and $z_0 = 0.5$. The reproductive rate of species z, $r_z = 3.7$. In A, $r_x = 1.6667$, in B, $r_x = 2.6667$. Stochastic variability over time of both species $r_t = 0.01$. Extinction of species x is indicated by hollow, of species z by grey, no extinction by blank. Note that a minor difference in alpha and beta can lead to a dramatic difference in the outcome of competition. Programs and algorithms for the programs can be found on www-personal.une.edu.au/~krohde/ (fuzzy/fuzzy chaos, 2 and 3 species competition).

the example of Darwin's finches on the Galapagos Islands in detail. Thus, morphological differences and particularly differences in the size of beaks of species inhabiting the various islands have been interpreted as being due to genetic drift, interspecific competition, recognition, and physical and vegetational differences between islands. Null-hypotheses (that species are independent of each other) appear to show that interspecific competition has little effect on birds of the Galapagos Islands and those of the New Hebrides, but the null-hypopthesis was rejected for birds and bats of the West Indies. However, the validity of the null-hypothesis used also has been rejected by some (for a detailed discussion and references see Connor and Simberloff 1986). Nevertheless, Connor and Simberloff conclude that, for evaluating non-experimental evidence, a null-hypothesis should usually be tested first. For some communities, predation may cause effects similar to those due to competition.

Density distributions were used by Janovy *et al.* (1992, 1995), who presented a lottery model, in which hosts acquire their parasites by Monte Carlo type sampling from multiple kind arrays. Assumptions for a null model were: "(1) successes are independent events and (2) it is possible to fail completely to acquire any parasites or their analogs." They postulated that "claims of parasite interaction need to be supported by studies showing departures from the predictions of this model." The authors state that their "model is applicable to any system that mimics a multiple-kind lottery" in which the above two assumptions can be made. They applied their model to data sets from three host/parasite systems in order to test for the occurrence of interspecific competition, and found that there were very few (2 out of 15) significant departures from the expected values. According to the authors, the two departures did not support the assumption that competition was responsible because they were the result of a greater and not a smaller than expected number of hosts infected with many species. The authors explained the departures by abiotic events. The study is of great value because it shows convincingly that the arguments given by various authors in support of interspecific competition are not valid. Nevertheless, the point can be made that, although a null model using density distributions calculated on the basis of certain assumptions (e.g., successes are independent events) can be established, such a model is biologically meaningless since successes very often are not independent and we cannot know a priori which ones are and which ones are not. As pointed out by Janovy *et al.* (1995), infective stages of parasites, for example, often use the same microenvironment, such as the same intermediate host, and the likelihood of becoming infected with more than one parasite species therefore may be much greater than random. In the cases discussed by Janovy *et al.*, the (perhaps remote) possibility exists that interspecific competition may indeed be common but remains undetected because expected values are overestimated. The section on "Larval trematodes in snails . . . " (pp. 131–134) indeed makes it clear that heterogeneity of the environment may lead either to a greater than expected likelihood of co-occurrence, or it may lead (less frequently) to a smaller than expected likelihood.

4 · *Interspecific competition: effects in communities and conclusion*

The previous chapter dealt with the effects of competition on species. Here we examine effects in communities, although it should be noted that there is some overlap: effects on species and communities cannot always be clearly distinguished.

Isolationist (individualistic, non-interactive) and interactive communities

Wiens (1984) distinguished interactive communities (structured by interactive processes, mainly competition), and non-interactive communities (communities largely "structured" by individualistic responses of species). He points out that most studies deal with interactive systems. Holmes and Price (1986) applied this distinction to parasite communities, distinguishing interactive and isolationist (non-interactive) infracommunities of parasites. In the former, colonization probabilities of hosts are high, and communities are likely to be saturated and equilibrial. In isolationist communities, probabilities of colonization are low, resulting in unsaturated, nonequilibrial communities. Holmes and Price, in their synthesis, conclude that distinguishing between isolationist and interactive communities is "probably too crude to be of lasting utility."

In the following section, I give some examples of interactive communities with evidence for interspecific competition, and of isolationist communities without such evidence.

Examples of competition in communities

Although evidence in many cases is poor, it seems nevertheless that competition is of some importance in many communities (for parasites see some contributions in Esch *et al.* 1990, and Lello *et al.* 2004).

Schoener (1983) reviewed evidence for interspecific competition in the past literature: in 90% of the studies and 76% of the species, some degree of competition was found. However, Connell (1983), who also reviewed such evidence, found indications of competition in not more than 43% of species. It must be said, however, that in none of the cases reviewed by Schoener and Connell were the strict criteria formulated by Connell (1980) actually applied. Underwood (1986) discussed evidence for interspecific competition from field experiments and concluded that in many communities competition is undoubtedly important. Lawton and MacGarvin (1986) drew attention to an important difference between insects and small herbivorous mammals. According to them, interspecific competition is not of great importance in the former, because numbers in populations are kept low by enemies (parasites, predators, disease), whereas in the latter competition is important (Schoener 1983), which implies that regulation by enemies is not effective. It has been suggested that large mammals, in some cases at least, are also subjected to much interspecific competition, which regulates their numbers (Sinclair 1979). However, other explanations (predation, resource limitation) have been given and are more likely (Sinclair 1985, further references in Lawton and MacGarvin 1986). Källander (1981) and Dunham (1980) found interspecific competition between species of tits and iguanid lizards, respectively, in some years but not in others. Of great significance, the careful and extensive studies of Kennedy (1985, 1992) on acanthocephalans of eel, *Anguilla anguilla*, in the British Isles suggest that even in species-poor isolationist communities interspecific interactions may be important, as evidenced by the observation that microhabitat width changed in the presence of another species, and that mixed species infections were very rare. However, in view of the great variability of distributions in single infections, the claim that concomitant infections lead to microhabitat changes needs verification, and the possibility cannot be entirely discounted that dominance of one species at a particular locality may be due to some environmental factor favouring one over the other species. Bates and Kennedy (1990) found one-sided interactions in high-level experimental infections of rainbow trout, *Oncorhynchus mykiss*, with two species of Acanthocephala. Lello *et al.* (2004), in an important study, used a combination of generalised linear modelling (GLM) and residual maximum likelihood (REML) linear mixed model analyses of parasite count data to test the null-hypothesis that parasite interactions do not influence parasite numbers in a free-ranging rabbit population in Scotland monitored over 23 years. They found consistent negative and positive interactions.

Examples of communities without evidence of competition

Rathcke (1976a,b), Strong *et al.* (1979), Strong (1981), and Price (1980) give examples of communities in which there is no evidence of competition (see also some contributions in Esch *et al.* 1990). For insects of *Pteridium aquilinum*, which are among the best studied communities (see Lawton 1999), and for other phytophagous insects (see Strong *et al.* 1984) there is little or no evidence of interspecific competition. Kennedy (1990) concluded that helminth communities of freshwater fish, on the whole, are chance and "isolationist" assemblages rather than structured communities, and Aho (1990) concluded that helminth communities of amphibians and reptiles are depauperate and non-interactive. According to Pence (1990), little is known about helminths of mammals but, although some mammals may have an interactive community of helminths (see Haukisalmi and Henttonen 1993a,b), in other mammalian hosts indirect evidence indicates that their helminth communities are isolationist and non-interactive (see also pp. 121–127). Combes (2001, pp. 411–413, references therein) briefly discusses a number of examples with clearly unsaturated infracommunities, i.e., isolationist communities of parasites with little or no evidence of interspecific competition. They include helminths of eel in Britain and Italy, trout, four species of the teleost *Lepomis*, Mediterranean sparid and labrid fishes, four species of salamanders, ectoparasites of Alaskan birds, some gull species, a swan in Poland, and bats. Some examples of interactive communities include intestinal helminths of lesser scaup (*Asythia affinis*) in Canada, and the coot *Fulica atra* in Poland. Combes gives more examples of isolationist than of interactive communities, and even some of the examples for interactive communities given by him are doubtful. Thus, he refers to a study of Kennedy and MacKinnon who have shown that two species of the nematode *Thelazia* "show distinct preferences in their locations within the eyes of cattle (eyelids, various glands, etc.), indicating that they also form an interactive community." No evidence that competition is responsible for the habitat segregation is given, which may also be due to random selection of habitats or reinforcement of reproductive barriers. (see pp. 39–47 and Chapter 5).

Some authors have used the asymptotic relationship between local and regional species richness as evidence for species saturation and the effects of interspecific competition, as discussed in the following.

Relationship between local and regional species diversity

Many authors have discussed how species comprising communities are acquired from the regional species pool through filters. At least three such filters can be distinguished, each of which leads to the loss of some species, as follows:

(1) large-scale biogeographic processes (distance, isolation);
(2) landscape filters (patch size, density, configuration);
(3) habitat availability (Lawton 1999, references therein).

But Lawton also points out that the importance of these filters may be secondary to that of local ones. Concerning the relationship between regional and local species richness, Cornell and Lawton (1992), also Lawton (1999, further references therein), following an approach introduced by Terborgh and Faaborg (1980), distinguish type I and type II systems, although in the real world systems may be intermediate. In the former there is proportional sampling, that is, local richness rises in proportion to regional richness. In the latter, local richness levels off with increasing richness, that is, it never reaches the richness of the regional species pool, because interactions between species prevent this. Both false type I and type II communities may seem to exist under certain conditions. For example, overestimation of regional species diversity may falsely lead to the conclusion that communities are of type II, and a pattern similar to type II may also result from "stochastic equilibrium" (colonization and extinction are in balance). Lawton (1999) briefly discusses a few examples and concludes that type I communities (or communities close to them) are more common than type II communities. Cornell and Karlson (1997), Srivastava (1999), and Lawton (2000, further references therein) reviewed the literature and also concluded that most systems appear to be type I communities. But even some of the examples for type II communities are probably not correct, as mentioned by Lawton (1999) himself. Thus, fish parasite communities almost certainly are not saturated, and interspecific competition is not important for them. Lawton, importantly, stresses that a pattern does not give a mechanism, i.e., other mechanisms than interspecific competition may lead to type II communities. For detailed examples on the relationship between local and regional diversity see pp. 130–131.

Rohde (1998b), using computer simulations, has shown that an asymptotic relationship between local and regional species richness may simply arise from different likelihoods of species occurring in a community

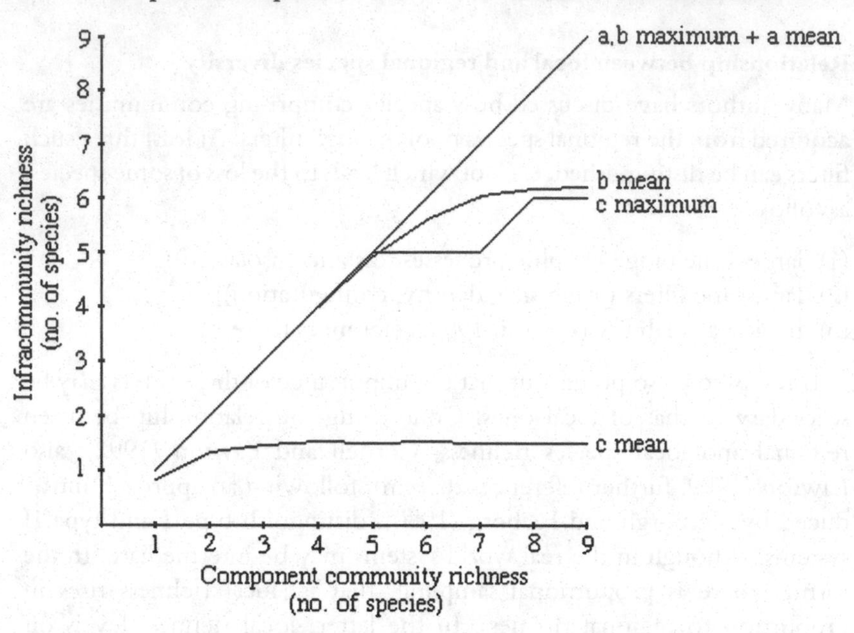

Figure 4.1. Computer simulations of the relationship between infra- and component community richness. The assumption is that each infracommunity can be recruited from any species in the component community, in any order, but species have different likelihoods of appearing in an infracommunity due to different transmission rates and intrinsic life spans. The richness of a community can vary between 1 and 9 species. 1000 iterations. (a) all species have a 100% likelihood of appearing in the infracommunity; (b) 7 species have an 80% likelihood, and 2 have a 30% likelihood of appearing in the infracommunity; (c) 2 species have a 40% likelihood, and 7 have a 10% likelihood of appearing. Note the asymptotic relationships for the means and maxima of all infracommunities except a maximum and mean and b maximum, although no assumptions on interspecific effects have been made. The conclusion must be that interspecific competition is not necessary to explain the asymptotic patterns. For access to program see www.personal.une.edu.au/~krohde/ (1) Infra- vs component community richness program (Macintosh only). Reprinted from Rohde (2002), with permission of Elsevier.

because of different life spans and colonization probabilities (Figure 4.1). But these results do not necessarily mean that the distinction of type I and II communities is wrong. What they mean is that, before the occurrence of type II communities due to competition can be assumed, results have to be tested against a null model in which species differ only in their likelihood of occurrence as a result of different life spans and colonization rates. But for how many systems have these parameters for all species been determined?

The demonstration that interspecific competition occasionally occurs in type II communities is not sufficient evidence that it limits local species richness. Another problem with distinguishing the two types of communities is that local and regional richness are statistically not independent (for references see Lawton 2000). Also, Rosenzweig and Ziv (1999) have shown that a linear relationship between local and regional richness cannot always be distinguished from a power curve. According to them, local versus regional richness patterns are "echoes" of species–area curves. These echoes are nearly linear, in fact so close to linear for certain parameters that they cannot be distinguished from linear although they are, in fact, power curves. Furthermore, Caswell and Cohen (1993) have shown that type I patterns can also arise in communities with strong competition, if patches of species are knocked out by environmental disturbances, and Godfray and Lawton (2001) presented a model in which type I patterns may exist even if competition limits species numbers. Interactive communities may have no limits to species numbers in certain models (e.g., Tilman 1999); Lawton interprets this as meaning that it is not interspecific competition per se that matters, but that competition must be sufficiently strong to limit species numbers. Finally, Shurin (2000) and Shurin *et al.* (2000) have recently demonstrated that communities of zooplankton exhibit linear relationships between local and regional diversity, but that such relationships are not incompatible with strong local interactions. Furthermore, such relationships are highly scale-dependent (Shurin and Srivastava, in press). All this shows that, as already pointed out by Lawton (see above), linear or curvilinear relationships between local and regional richness do not say anything about the mechanisms responsible for the patterns.

Many authors have used the distinction between the two types of community to test for saturation (e.g., Cornell 1985a, b; Aho 1990; Tonn *et al.* 1990; Hawkins and Compton 1992; Kennedy and Guégan 1994, 1996; Oberdorff *et al.* 1998), but a null model as discussed above has not been used in any of these studies. Also and importantly, as pointed out above, several studies have shown that a linear relationship between local and regional species richness does not exclude the possibility of saturation per se, and Shurin and Srivastava (in press), evaluating empirical evidence from the literature, believe that – in spite of the predominance of linear relationships – the overall picture seems to be more consistent with saturation. However, it is important to check these assertions using other evidence (not related to local versus regional diversity patterns; see pp. 39–48).

Pool exhaustion may explain the asymptotic relationship between local and regional diversity, even in the absence of competition, as pointed out by Lawton and MacGarvin (1986) who discussed the observation that numbers of species of insects on introduced plants at first increase rapidly and then more and more slowly, until a maximum is reached. They consider the possibility that interspecific competition is responsible as unlikely, because these plants have many empty niches. A more likely explanation is that species for successful colonization are not available (pool exhaustion hypothesis of Lawton and Strong 1981). This may be an important mechanism explaining poor communities.

The packing rules based on fractal geometry and competition

Ritchie and Olff (1999) have used spatial scaling laws (fractal geometry) to derive a rule for the minimum similarity in the size of species that share resources. Serengeti (East African) mammalian herbivores and Minnesota savanna plants were shown to conform to the predictions of the packing rules (pp. 41–42). However, Rohde (2001a) has shown that the packing rules do not apply to metazoan ecto- and endoparasites of marine fishes (pp. 43–44). These negative results support the view that parasites of marine fish do not live in saturated structured communities but rather in assemblages not significantly structured by interspecific competition. The positive results presented by Ritchie and Olff (1999) for Serengeti grazing mammals and North American savanna can be explained by the fact that they are either vagile (mammals) or disperse well (savanna plants), and both utilize significant proportions of the resources for which different species compete, plants in the case of the former and light and space in the case of the latter. Interspecific competition is therefore expected (for further details see pp. 178–180).

Even if competition occurs and even if it is important in structuring communities, effects on infra- and component communities may be quite different, as illustrated by the example of trematode communities in snails. Infracommunities of parasites are defined as all populations of all species infecting a single host individual, component communities are all the populations of all parasite species infecting a host population.

Different effects of competition in infra- and component communities

Of great importance, the studies of Sousa (1990, 1992, 1993) and Kuris (1990) on the community structure of larval trematodes (18 species) of the coastal snail species *Cerithidea californica*, at a number of sites on the Californian coast, led to the conclusion that trematode species interact at the infracommunity level. This conclusion was based on the findings that mixed (multi-species) infections were fewer than expected under the null-hypothesis (trematodes are randomly and independently distributed), that in the few multiple infections rediae of one species were observed to feed on rediae, sporocysts, and cercariae of another in a hierarchical order, and that in marked snails with known infections, parasite species at the bottom of the hierarchy were replaced over time by species higher in the hierarchy. The first point, on its own, would not be convincing, since infections may be acquired in a very heterogeneous fashion and a null-hypothesis can therefore not be clearly defined, but joint evidence from all three points is convincing. However, some of the observed interactions were in fact predatory rather than competitive in the strict sense: rediae feeding on other larvae.

In spite of the clear evidence for interspecific competition (plus predation) at the infracommunity level, evidence from observations at the component community level, that is, not within a host individual but in a host population, did not support the view that interspecific competition has significant effects. This conclusion is based on the findings that species richness and diversity of trematodes increased with snail size, i.e., complete dominance by a few species did not develop and species accumulated with time, and that neither numbers of uninfected hosts nor variation in host size was correlated with parasite diversity; a greater proportion of older than younger snails were infected. Sousa (1990) stressed that the findings did not exclude the possibility that there were some interspecific interactions, but any reductions due to competition were "more than compensated for by increases in both the number and equitability of other parasite species in older host populations." The findings on snail–trematode systems are of particular interest because of the large size of the parasites relative to the snails. On these grounds alone, interspecific effects could be expected (for a detailed discussion of trematode communities in snails see pp. 131–134). Some parasites of birds are supposed to have saturated component communities but resources at the infracommunity level are underutilized (Bush 1990).

Competition in mature and young communities

Holmes (1973) held the view that helminth communities are mature and to a large degree structured by biotic interactions, i.e., competition, apparently in the past, because most species exhibit "selective site segregation of niches" (utilization of different sites that prevents interspecific competition), and only few exhibit interactive site segregation, where sites are reduced in size when competitors are present. Price (1980) holds a contradicting view, that parasite communites are, in fact, not mature and not significantly shaped by past competition. Competition, where it occurs, is intermittent. However, age does not necessarily imply structure by competition leading to selective site segregation. Platyhelminthes are a very ancient group and they contribute a significant proportion of species to parasite and other communities, but evidence for competition now or in the past is poor (see Chapter 10).

Competition in saturated but less so in unsaturated communities

Rohde (1980a) generalized his findings on ectoparasites of marine fish to include groups of organisms other than parasites, and postulated that not only parasites, many if not most of which have small population sizes, but free-living species that are also rare, live under conditions where not interspecific competition but selection to enhance mating opportunities and reinforcement of reproductive barriers are important. However, for large animals with great vagility, particularly birds and mammals, and free-living insects with great vagility and large populations that can rapidly spread into "vacant niches," competition is probably important. He further hypothesized that, for the vast majority of species, competition does not reduce niches in the course of evolution, but there will be a gradual filling of niche space, partly by sympatric speciation, and by separation of overlapping niches in order to avoid hybridization. Niches do not expand into empty niche space, because the reproductive capacity of most species is too low to guarantee mating and thus cross fertilization in suboptimal niches. The hypothesis was tested by Gotelli and Rohde (2002) using null-model analyses. They found that communities of large animals (birds, mammals) are highly structured, whereas communities of herps (amphibians and reptiles) and fish parasites are not. For a fuller discussion see Chapter 11.

General aspects and conclusion

Competition does not decide the distribution and abundance of organisms

White (1993) emphasizes the importance of seeing that selection is a negative process, i.e., the unfit are selected against. Nitrogen is the most limiting chemical for plants, and competition is a consequence of and not a cause of its limitation. In other words, "competition does not decide the distribution and abundance of organisms. That is decided by the inadequacy of the environment." There is growing evidence that interspecific competition, although it occurs, and in contrast to intraspecific competition, is relatively uncommon, and of debatable significance for evolution. He gives many examples of the non-existence of competition between coexisting species, explained by the fact that their environments are different, i.e., they use different niches. He discusses three examples from the literature in more detail (for references see White 1993), that of the Australian smoky mouse *Pseudomys fumeus* and three other small mammals seemingly sharing the same habitat, that of four species of ground finches on the Galapagos Islands, and that of a gall-forming aphid and a leaf-eating beetle. In each case, species seemed to share the same habitat but closer examination revealed differences in microhabitat and/or food use. Coexisting species were not excluded by competition from particular parts of the habitat/food resources, but differed because of the distribution and abundance of their respective foods. Emphasis on competition deflects attention from what is ecologically significant for a species. White concludes that there can be no doubt about the occurrence of interspecific competition, because it "is a real and potent cause of the direction in which many organisms evolve" but it is not "necessary." The inadequate environment performs the selecting out of unfit organisms, even if competing species are not present. This implies nonsaturation of habitats, empty niches, although this is not explicitly stated by White. Concerning the existence of self-regulating mechanisms as, for example, brought about by social behavior (e.g., Wynne-Edwards 1962), or by changes in the proportion of "inferior" and "superior" genotypes in a population, White maintains that they do not exist.

The occurrence and significance of interspecific competition: a conclusion

The discussions in this chapter and in Chapter 3 have shown that much of the evidence for interspecific competition is faulty. A major problem in demonstrating the effects of interspecific competition is the difficulty in formulating valid null-hypotheses. Nevertheless, interspecific competition is likely to occur in certain taxa and under certain conditions (see Chapter 11). But even where it occurs, the outcome of competition may be largely unpredictable (see pp. 65–67). The uncertain outcome of interspecific competition due to chaos, environmental stochasticity or other factors, reduces the likelihood that it has significant evolutionary effects. Overall, occurrence of competition is not evidence for its evolutionary significance. It seems that, on an evolutionary scale, interspecific competition does not play the major role assumed by many, as evidenced by the rarity of cases in which divergence due to competition has been made likely.

5 · *Noncompetitive mechanisms responsible for niche restriction and segregation*

Niche restriction is usually attributed to interspecific competition: a species does not spread from its optimal niche into less suitable ones because competing species, better adapted to these niches, prevent it from doing so. However, the chapters dealing with interspecific competition have shown that evidence for this assumption is in many cases far less convincing than generally assumed. But why, then, are niches restricted? This chapter deals with this problem by firstly demonstrating that niches may be restricted even if potentially interacting species are absent, and secondly by suggesting other, non-competitive mechanisms responsible for niche restriction. It further proposes alternative mechanisms for explaining niche segregation.

Evidence for niche restriction even in the absence of potentially interacting species, and mechanisms responsible

Gill parasites of fishes are particularly good models for examining this question because the distribution of parasites can be easily mapped, because the number of species is limited (even in the richest communities they do not exceed about 30 species and are usually much fewer), and because an almost unlimited number of replicas are available. Rohde, in a series of papers (e.g., Rohde 1976a, b; 1977a, b, c; 1978a, b; 1980b; and reviews 1989; 1991; 1994a; 2002), has shown that monogenean gill parasites of marine fish from all latitudes use microhabitats that are sometimes very restricted, even when competing species do not exist or, when they do exist, are not present on individual fish. Figure 5.1 gives an example. Six fish were examined and found to have one species of Monogenea on the gill filaments (maximum 18 worms on one fish), all of which chose the same microhabitat, the base of the gill filaments in the middle portion of the gills. Only two individuals of other parasites were present, one copepod larva on the gill filaments and one adult copepod in

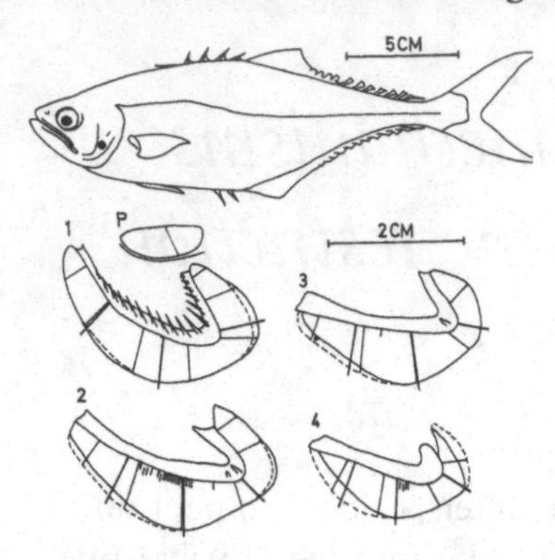

Figure 5.1. Distribution of one species of Monogenea (lines), one adult copepod (large filled circle), and one larval copepod (small filled circle) on the gills and in the mouth cavity of six *Oligoplites palometa* off the coast of Brazil. Note: the habitat is largely empty (related fish species were found to have much larger numbers of species occurring at greater abundances). In spite of this, species have narrowly defined microhabitats. From Rohde (1991). Reprinted by permission of the editor of Oikos.

the mouth cavity, suggesting that niche restriction is not the result of interspecific effects, but genetically programmed. The proximate causes for niche finding in this case have not been studied, but may be water currents, size of the gill filaments, or some other factors. There is now ample evidence that preference for microhabitats is, at least to a large degree, not affected by the presence of other species: even species that always occur singly may have very narrow microhabitats, although environmental factors such as oxygen concentration, or crowding (i.e., presence of large numbers of individuals of the same species) may have effects.

Two non-competitive mechanisms for niche restriction have been proposed, the necessity to specialize in order to guarantee survival in a particular microhabitat, and mate finding. Niche space is largely empty, as shown in Chapter 2 ("Species nonsaturation and nonequilibria"). All species must be specialized to varying degrees, and this alone will lead to niche restriction even in largely empty niche space. Price (1980), in particular, emphasized this aspect of niche restriction. Rohde (1976a, b; 1977b) suggested that one ultimate reason (i.e., one biological function) for narrow niches may be the enhancement of mating encounters.

Evidence for the "mating hypothesis of niche restriction" was again provided by using fish and other parasites. Many of these parasites are hermaphroditic (all Monogenea, most Digenea, all Aspidogastrea, all Gyrocotylidea, all Amphilinidea, the Eucestoda). Therefore, it is important to point out that these species often show obligatory cross-fertilization.

Thus, Rohde (1973) found that the aspidogastrid *Lobatostoma manteri*, in single infections, produced only haploid eggs which did not develop beyond the early blastula. Kearn (1970) found no evidence for self-fertilization in the monogenean *Entobdella soleae*, and, although the first daughter animal of the monogenean *Gyrodactylus alexanderi* is produced without cross-fertilization, this may then occur in the production of the second embryo (Lester and Adams, 1974). The importance of cross- and self-fertilization have been particularly well studied in digenean trematodes (Nollen 1993, further references therein). In the experiments described by Nollen, one species never self-fertilized even in single infections, others self-fertilized only when isolated, and others occasionally self-fertilized even in multiple infections. All species examined cross-inseminated when kept in groups. Mature worms of one species when transplanted singly, stopped growing and their reproductive system rapidly degenerated. Even in self-fertilizers, it may well be that cross-fertilization is necessary in the long run, that is, numbers and/or viability of offspring may be reduced if cross-fertilization does not occur at least occasionally. Studies over many generations are necessary to confirm this. Indirect but very strong evidence for the importance of cross-fertilization in some hermaphroditic helminths is the amazing complexity of copulatory organs in many monogeneans. It would be difficult to understand why such a complexity should evolve in species without cross-fertilization. Many parasites, of course, are bisexual (copepods, almost all nematodes, isopods, branchiurans, etc.) and the necessity of cross-insemination is self-evident for these. In some species, the factors involved in mate finding have been elucidated. Thus, Fried, Tancer and Fleming (1980) studied pairing of the digenean *Echinostoma trivolvis* in vitro and found that free sterols were the main chemoattractants (Fried and Diaz 1987).

Rohde (1976a,b; 1977b; 1979a; 1980a; 1984; 1989) gave the following evidence for the "mating hypothesis" of niche restriction for parasites:

(1) narrow host ranges and microhabitats lead to increased intraspecific contact;
(2) adult stages often have fewer hosts and narrower microhabitats than sexually immature and larval stages;
(3) microhabitats of sessile and rare species are often narrower than those of more motile and common species; and
(4) microhabitats of some species were shown to become more restricted at the time of mating.

The first point is self-evident, and Rohde (1977b) has given some examples, i.e. metazoan ectoparasites on the gills of *Scomberomorus commerson*. Five species of Monogenea and one species of copepod, all with distinctly restricted microhabitats, showed far greater intraspecific than interspecific contact. Experimental evidence for the effectiveness of microhabitat restriction on mating success comes from the study of Nollen (1993) on cross-insemination rates in the two digenean trematodes *Echinostoma trivolvis* and *E. caproni*. The first species occurs throughout the jejunum and ileum of experimentally infected hamsters, the second congregates in one or two areas of the ileum in moderate infections in mice and hamsters. The first species had a cross-insemination rate of 14%, the second 52%.

Concerning the second point, as a rule, many metacercariae (larval trematodes) are found widely spread in muscle and other tissues of fishes (although there are some species with very narrow sites), whereas adult flukes usually occupy restricted microhabitats in the digestive tract. For example, *Stephanostomum baccatum* lives in the rectum and lower intestine of several marine fish species, while its metacercariae infect various somatic muscles, the operculum, fins, pericardium, liver and mesentery of fish (references in Rohde 1989). Likewise, protozoan cysts are usually spread over all gills of infected fish (Rohde 1989).

Rohde (1980b, 1989) gave several examples for the third point. Thus, the sessile copepods *Neobrachiella impudica* and *N. bispinosa* on *Trigla lucerna* in the North Sea occupy much smaller microhabitats than the mobile copepods *Caligus diaphanus* and *C. brevicaudatus* on the same host. A small monogenean of the subfamily Ancyrocephalinae, occurring in large numbers, infects all parts of all gills of *Hyporhamphus quoyi* in New Guinea, whereas rare species on other fish in the same region have distinctly restricted microhabitats on the gills. However, as pointed out by Rohde (1989), the force of this argument is somewhat reduced since some species have larger microhabitats when found in large numbers than when they are scarce (e.g., Arme and Halton 1974; Ramasamy *et al.* 1985), although at least some gill parasites do not show such microhabitat expansion at high infection intensities (Rohde 1991).

Kamegai (1986) has provided evidence for the fourth point. He demonstrated that juveniles (diporpae) of the monogenean *Diplozoon nipponicum* on *Carassius carassius auratus* are found on all gills until the fourth day after infection. Then, they gather on one gill and copulate for life. Lambert and Maillard (1974, 1975) also showed smaller microhabitats of adults of two species of the monogenean genus *Diplectanum*, apparently as the result of migration of the juveniles. According to Kearn (1988), the monogenean *Entobdella soleae*, after transfer to another host individual, the sole *Solea solea*, migrates from the upper to the lower surface via the head; some transferred individuals, particularly adults, spend longer than others on the head, thereby improving the chances of cross-fertilization between worms arriving on the host at different times. Another example is the parasitic prosobranch snail *Thyca crystallina*. Juveniles settle largely on the upper surface of the distal parts of the arms of the sea star *Linckia laevigata*, where they are randomly orientated, whereas large females face the starfish mouth in a small area on the right side of the ambulacral groove on the surface of the oral arm (Elder 1979).

The mating hypothesis of niche restriction is not limited to parasites, as indicated by the observations of Zwölfer (1974a,b) on trypetid flies, most of which use certain host plants as a meeting place for mating. And this phenomenon is indeed common in insects. According to Zwölfer (personal communication, and examples in Zwölfer and Bush 1984), the "Treffpunkt" or "rendezvous principle" applies to almost all host-specific phytophagous insect species (Diptera, Coleoptera, Hymenoptera, and many Lepidoptera). Males of the species use and defend the meeting places as temporary, resource-based territories. In species with endophytic larvae such meeting places are the rule.

Niche segregation to ensure reinforcement of reproductive barriers

Interspecific competition has not only been proposed as the principal cause of niche restriction, but also of niche segregation. However, as in the case of niche restriction, there are other alternatives to explaining niche segregation by interspecific competition. An important alternative explanation (beside random selection of niches in largely empty niche space, which was discussed on pp. 39–48) is reinforcement of reproductive barriers. Miller (1967) has already stressed that two explanations for character displacement are possible, i.e., competition ("ecological character displacement"), or reinforcement of reproductive barriers ("Wallace effect"). Several authors (e.g., Grant 1972; Wilson 1975; Connell 1980; Arthur 1982) have pointed out that the ecological aspect of character displacement is weak in many situations. Concerning ectoparasites of fish, Rohde (1989) concluded that there is little or no evidence for the view

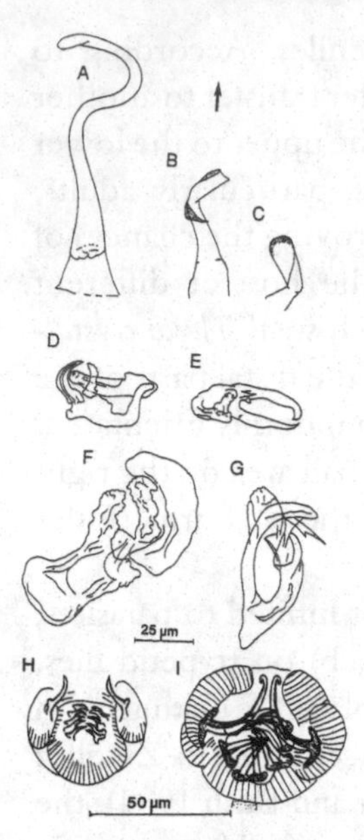

Figure 5.2. Monogeneans infecting the gills of marine fish may use identical microhabitats on the gills when they differ in the shape (and/or size) of their copulatory organs; they are always segregated when they have identical copulatory organs, suggesting that not interspecific competition but reproductive segregation is the primary reason for species segregation. The following examples illustrate this: they show copulatory sclerites of monopisthocotylean monogeneans infecting the gills of A. *Lethrinus miniatus*, B. *Haliotrema lethrini*, C. *H. fleti*, D. *H. chrysostomi*, E. *Calydiscoides gussevi*, F. *Protolamellodiscus* sp., G. *Calydiscoides difficilis*, *C. autralis*, and of polyopisthocotytleran monogeneans on the gills of *Scomber* spp., H. four species with copulatory organ, and I one with copulatory organ. All species on *L.miniatus* except for E inhabit identical or overlapping microhabitats, the four species H on *Scomber* spp. are segregated in different microhabitats or hosts, species I overlaps with some of the others. Arrow points to anterior end. Reprinted from Rohde, Hayward, Heap, and Gosper (1994), with permission of Elsevier.

that interspecific competition has been an important evolutionary force leading to niche segregation, but that there is much support for the view that mechanisms leading to reproductive isolation have been important.

Evidence for reinforcement of reproductive barriers as an agent of niche segregation in parasites was discussed by Rohde and Hobbs (1986) and Rohde (1989, 1991). Rohde and Hobbs (1986) compared niche overlap in 35 parasite species on the gills and in the mouth cavity of six species of marine fish. Nineteen species occurred in congeneric pairs or triplets. They found, using a newly proposed asymmetric percent similarity index (Rohde and Hobbs 1986, 1999), that congeners overlap less than non-congeners, and that those congeners that showed considerable overlap had markedly different copulatory organs, in contrast to those that were spatially segregated. The only limiting factor for all of these parasites is space for attachment, since food (blood, epithelial cells and mucus) are in unlimited supply as long as the host is alive. But use of space does not differ between congeners and non-congeners. The authors concluded that not competition for space, but reinforcement of reproductive barriers is responsible for niche segregation in those cases in which it occurs. As phrased by Rohde (1991) "if competition to reduce resource exploitation was responsible for segregation, it should affect species with similar or dissimilar copulatory organs in the same way" (Figure 5.2). For a recent, very well documented study of niche segregation in 52 monogeneans of the genus *Dactylogyrus* infecting 17 species of freshwater cyprinid fish see Jarkovsky *et al.* (2004, further references therein), who showed that not only differences in the size and shape of copulatory organs, but also of attachment organs are important for reproductive segregation.

One of the most convincing studies of character displacement which has led to reinforcement of reproductive barriers is by Kawano (2002), who examined male morphology of two closely related rhinoceros beetles, *Chalcosoma caucasus* and *C. atlas* in Laos, Thailand, Malaysia, Indonesia and Mindanao. The beetles occur in 7 sympatric and 12 allopatric locations. The species has major and minor morph males. Major morph males have large horns and fight with other major morph males, whereas the smaller minor morph males, which have rudimentary horns and relatively large wings, secure mating by emerging earlier in the year and the day and by flying over longer distances, thus avoiding contact with major morph males. The qualitative features and the variation in each species is the same in allopatric and sympatric locations, and there is almost complete overlap in dimorphism, body size, horn size, and size of genitalia between the two species in the allopatric locations. In all sympatric locations, differences between species in all characters are highly significant (Figure 5.3). In particular, differences in the size of genitalia are much greater than would be expected if due to general body size displacement (Figure 5.4). The author concludes that the differences have evolved to avoid interspecific competition and bring about reproductive isolation. However, evidence for interspecific competition is fairly weak if there is evidence at all, but there can be no doubt (in view of the much greater size differences of genitalia than of body length in sympatric

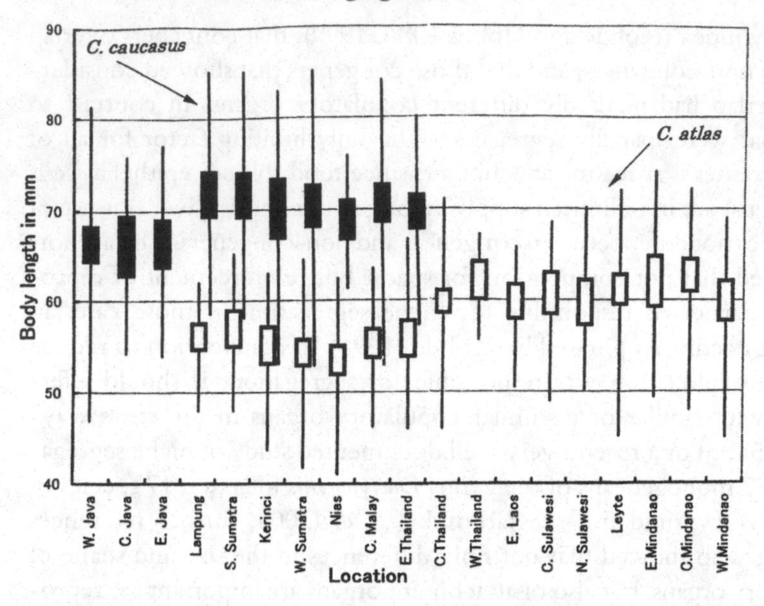

Figure 5.3. Body length of two rhinoceros beetles, *Chalcosoma caucasus* and *C. atlas* in allopatric and sympatric locations. Boxes show 5% reliable range of mean of each population, vertical lines show maximum and minimum values. Note the significant differences between the two species in all sympatric locations. From Kawano (2002). Reprinted by permission of the author and the University of Chicago Press.

locations) that reinforcement of reproductive barriers is an important factor, and perhaps the only factor involved. The same explanation may apply to the observation that one species occurs at higher and the other at lower altitudes in sympatric than in allopatric locations. It is unclear how the two species have evolved. Kawano discusses three possibilities: (1) species have evolved by allopatric speciation and have come into secondary contact in some locations; (2) species have evolved sympatrically and in some locations remained sympatric, whereas in others became allopatric; (3) the two species in the sympatric locations may be phylogenetically closer to each other than to the same species in other locations, i.e., *C. caucasus* and *C. atlas* have evolved repeatedly in each sympatric location. However, the third explanation seems unlikely in view of the existence of not only one but several sympatric "species" pairs.

Our conclusion for this chapter is that alternative explanations to both niche restriction and segregation by interspecific competition are possible, and are indeed in many cases more likely than competition. For

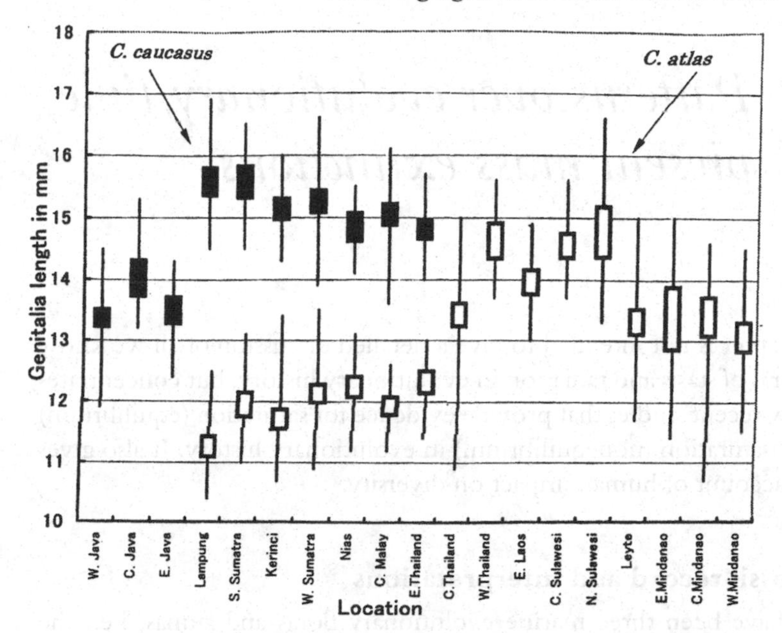

Figure 5.4. Length of genitalia of two rhinoceros beetles, *Chalcosoma caucasus* and *C. atlas* in allopatric and sympatric locations. Boxes show 5% reliable range of mean of each population, vertical lines show maximum and minimum values. Note that the differences between the two species in the sympatric locations are much greater than those for body size illustrated in Figure 5.3, suggesting that reproductive segregation may be the primary factor responsible for the differences. From Kawano (2002). Reprinted by permission of the author and the University of Chicago Press.

niche restriction these explanations are the need to specialize and to enhance mating encounters, and for niche segregation they are random selection of niches in largely empty niche space and reinforcement of reproductive barriers.

6 · *Patterns over evolutionary time, present mass extinctions*

This chapter is not intended to give a detailed discussion of all we know about eras of stasis and radiation in evolutionary history, but concentrates on a few recent studies that provide evidence for saturation (equilibrium) and nonsaturation (nonequilibrium) in evolutionary history. It also gives a brief account of human impact on diversity.

The fossil record and interpretations

There have been three marine evolutionary floras and faunas, i.e., the Cambrian, Paleozoic and Modern, each with its own degree of diversity, and each subsequent one with higher diversity than the previous one (Jablonski and Sepkoski 1996). Benton (1995, 1998), re-analysing fossil evidence, has shown that there has been an exponential increase in the number of families of continental and marine organisms in geological time to the Recent. Courtillot and Gaudemer (1996) analysed the same data, and arrived at the somewhat different conclusion that equilibria were reached several times but re-established at higher levels after each mass extinction; but they still found an increase over geological time. Jablonski (1999, see also Jackson and Johnston 2001) has shown that fossil data are solid and that the general trend of increasing diversity has not been changed by more recent data collected between 1982 and 1992 (Figure 6.1): there was a sharp rise in diversity in the Cambrian, a Paleozoic plateau interrupted by several mass extinctions, and a sharp rise since the Triassic, also interrupted by several extinction events. Thus, the fossil findings lend strong support to the view that saturation, overall, has not been reached.

However, the apparently overwhelming evidence for a substantial and even exponential increase in diversity of taxa over much of geologic time has been doubted. Rosenzweig (1995, references therein) believes that the increases as suggested by the fossil record over much of the time span are, to a large degree, not real. An important error is introduced by the

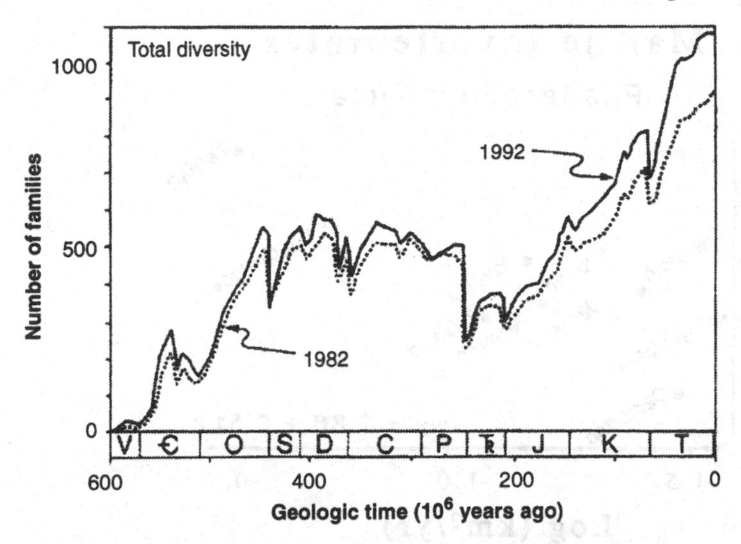

Figure 6.1. Fossil record of marine diversity (data from 1982 dotted line, data from 1992 uninterrupted line). Note several mass extinctions, Paleozoic plateau and post-Paleozoic rise in diversity. V - Vendian, Є - Cambrian, O - Ordovician, S - Silurian, D - Devonian, C - Carboniferous, P - Permian, Ŧ - Triassic, J - Jurassic, K - Cretaceous, T - Tertiary. Recent studies have not led to a significant change of trends, suggesting that the records indicating a continuing increase in diversity are sound. From Jablonski (1999). Reprinted by permission of the author and the American Association for the Advancement of Science.

effects of sampling areas, which increase towards the Recent (Figure 6.2). Those increases that did occur were restricted to short periods; for periods of tens of millions of years diversity changed very little. For example, diversity of Ordovician invertebrate muddy benthos remained more or less stable over five million years, when correction is made using "closed-model jack knife estimates" (for details see Rosenzweig) (Figure 6.3). But this stability was dynamic, species became extinct and were replaced by others. Likewise, North American large mammals retained their diversity over much of the Cenozoic.

A repeated decrease in diversity due to mass extinctions is well documented (see above and Jablonski 1991), and the removal of provincial barriers has also reduced diversity, although less dramatically, at least for higher taxa. Thus, before South and North America joined, the total number of mammalian families was 50 (25 families in North America, 23 in South America, and 2 in both continents), whereas it is now 38 (8 endemic northern, 15 endemic southern, and 15 shared families) (Flessa 1975). There have also been periods of prolonged stasis, as already

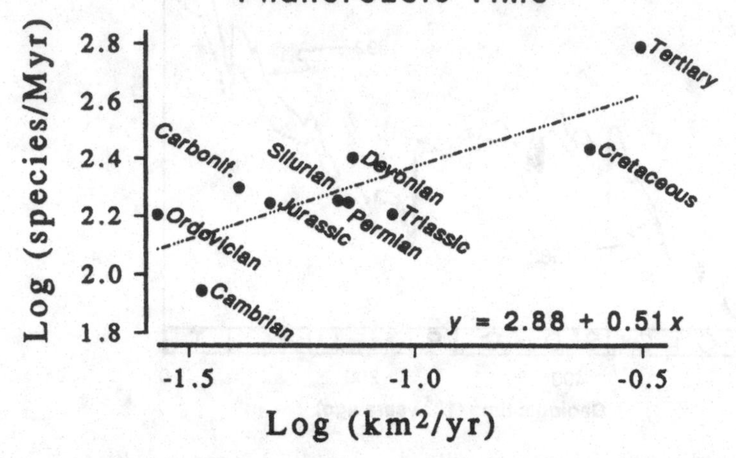

Figure 6.2. Species diversity of marine invertebrates correlated with rock area available for sampling. Rosenzweig interprets these findings as evidence that the apparent increase in diversity over geologic time is in fact an area effect. Data from Raup and Sepkoski after Rosenzweig (1995). From Rosenzweig (1995). Reprinted by permission of the author and Cambridge University Press.

mentioned (see, e.g., Benton and Pearson 2001). Rosenzweig's (1995) emphasis is on these steady states and on equilibrium in the past and now. He admits that diversity rose somewhat over hundreds of millions of years. One of the reasons for this is colonization of new habitats, such as the muddy ocean bottom. There may also be an increase in "versatility" leading to diversification, for which, however, there is no definitive evidence, but overall, these increases are thought to be relatively minor.

However, the objections of Rosenzweig and others can be rejected on the basis of a recent study by Jablonski *et al.* (2003), who shows that the so-called "pull of the Recent", i.e., possibly inflated diversity estimates resulting from sampling bias (more complete sampling of Recent biota) account for only 5% of the Cenozoic increase in bivalve diversity; 906 of 958 living genera and subgenera have existed in the Pliocene and Pleistocene. Bivalves represent a major component of the marine record, and it can be safely assumed that there has been a strong increase in diversity throughout the Cenozoic.

How has diversity of parasites and other "dependent" species, such as symbionts or commensals, evolved? Parasites comprise a very high proportion of all animal species. One quarter of the 40 000 animal species in

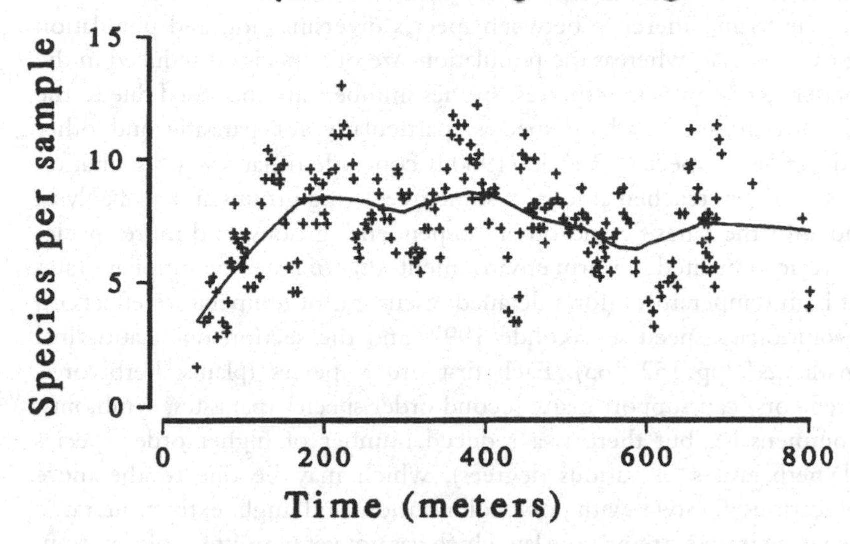

Figure 6.3. Diversity (species per sample) of Ordovician marine benthos. Note rise in diversity in the late Ordovician, and diversity of 6–8 species per sample during the following 4 Myr. These data suggest that there has been no increase in diversity throughout the Ordovician. Time measured as thickness of geological layer. From Rosenzweig (1995). Reprinted by permission of the author and Cambridge University Press.

Germany are parasites, not including parasites of plants (Arndt 1940), and more than half of the British fauna were estimated to be parasitic, including parasites of plants (Price 1977). Since very few fossil parasites are known (for a list see Combes 2001), statements on their evolution must be based on inferences from studies of extant species.

The opportunities provided by free-living animals to parasites (and hyperparasites, predators, symbionts, commensals) were taken into account in the diversification equation developed by Rohde (1980d) and based on the logistic equation for population growth. It is:

$$\frac{dS(t)}{dt} = (b(t) - e(t))S_f(t)\frac{K_f(t) - S_f + \sum_j^n a_{ij}(t)S_j(t - T)}{K_f(t)},$$

where dS/dt = rate of species diversification, b = rate of species formation, e = rate of species extinction, S_f = number of free-living species, K_f = carrying capacity for free-living species, $\Sigma a_{ij}S_j$ = the sum of all opportunities provided by species j for the formation of new species i, $(t - T)$ = time lag in species diversification, (t) = factor indicating variability due to environmental effects, and n = total number of species.

The main difference between species diversification and population growth is that, whereas the population size of a species is reduced in the presence of competing species, species numbers are increased due to the opportunities created by species, particularly for parasitic and other "dependent" species. Rohde (1980d) concluded that species saturation has not been reached at least in some free-living groups and probably in none of the parasitic and other "dependent" groups, and more species have accumulated in warm environments due to faster evolutionary rates at high temperatures (for a detailed discussion of temperature effects on evolutionary speed see Rohde 1992, and the section on "Latitudinal gradients", pp. 152–165). Each first order species (plants, herbivores, predators) can support many second order species (parasites, symbionts, commensals), but there is a reduced number of higher order species (hyperparasites of various degrees), which may be due to the more precarious lifestyle with low colonization and high extinction rates, resulting in an extreme time lag which has not yet permitted colonization. Evolution is slower in cold environments (e.g., Rohde 1992). Therefore, saturation of habitats with species would be reduced at high latitudes, with greater relative species diversity (number of species per host species) of some parasite groups in tropical than in cold-temperate seas, and in the Pacific than in the Atlantic Ocean. There would be diffusion of many tropical species into cold-temperate regions, and occurrence of many "progressive" traits in the tropics. However, exceptions are possible. Thus, it seems possible that in certain situations, e.g., the great East African lakes with their large numbers of recently evolved cichlid fish species, parasites could not keep up with the diversification of their hosts, which may have led to reduced relative species richness (number of parasite species per host species) of the parasites. Also, paradoxically, in spite of the higher saturation with free-living organisms at low latitudes, the number of extant vacant niches for "dependent" species, such as parasites or symbionts, may be even greater in the tropics, because of the larger number of potential host species.

Future studies should aim towards getting estimates for numbers of "secondary" species such as symbionts, parasites, etc., and assessing their contribution to the overall diversity. Estimates of local and global species richness generally ignore them, but they are of overriding importance not only for estimating current diversity but also for estimating numbers of species lost or endangered due to human impact, as will be shown in the following.

Recent and present extinctions

The discussion above shows that there has been a significant increase in diversity over evolutionary time, although there were periods of stasis or even mass extinctions leading to reduced diversity. All available evidence indicates that a new era of species collapse began thousands of years ago as the result of the evolution and spread of man on Earth, and that this collapse is continuing at an accelerating rate. Freeman and Herron (2004) have given an illuminating account of what is happening, underpinned by many quantitative examples from various sources. They refer to Steadman (1995) who estimated that 2000 species of birds have become extinct in the Pacific region over the last 2000 years. In New Zealand, to mention just one example, 44 bird species disappeared after the arrival of man. The extant fauna of the sheep rangelands of Australia includes about 30 species of amphibians, 220 species of reptiles, 200 species of birds, and 80 species of (mainly small nocturnal) mammals (Robertson *et al.* 1987). However, the fossil record indicates that about 30 000 years ago, before or at the beginning of human settlement, diversity of mammals at least was much richer and included much larger species, such as many large kangaroos, a rhinoceros-like marsupial, a large wombat, a large koala, all of them now extinct. When modern Europeans moved into the rangelands of western New South Wales, 16 of 22 marsupials and 5 of 7 rodents disappeared. Freeman and Herron (2004) refer to data compiled by Smith *et al.* (1993), who list the number of species from various major groups that have become extinct in the last 400 years, and those that are threatened by extinction. The list includes molluscs, crustaceans, insects, vertebrates, and among plants gymnosperms, dicotyledons, monocotyledons, and palms. A total of 0.3% of plant species are already extinct, and 9% are threatened, if current trends continue. However, the authors caution that only estimates for birds and mammals are likely to be approximately correct, whereas those for insects and plants are almost certainly underestimates. If current trends are enhanced, which is likely in view of the increasing effects of global warming, a much higher proportion of species will be threatened. There can be no doubt that global warming has occurred, although estimates on the degree of warming to be expected in the future and its effects on species extinctions vary widely. Hughes (2003) discusses climate change in Australia, including present trends, projections, and impacts on diversity. Australia, similar to trends elsewhere on Earth, has warmed by about 0.8 °C over the last century. Projections are that, by 2030 and 2070, annual average temperatures will increase (relative to 1990) by 0.4–2.0 °C, and 1.0–6.0 °C, respectively. Predictions for changes in rainfall are even more difficult, although increases in potential evaporation and reductions in the extent and duration of snowfall over much of the continent are likely. Hughes discusses the effects of climate change on various ecosystems. We select one example, i.e., coral reefs, which are of particular concern to Australia because the Great Barrier Reef is the largest system of coral reefs on Earth. In 1998 alone, approximately 16% of the living corals on Earth died, and in the Indian Ocean mortality was more than 40%. The main culprit was coral bleaching, i.e., the loss of symbiotic zooxanthellae essential for the survival of corals, due to increased temperatures. Modelling showed that bleaching comparable in strength to that in 1998 will become common within 20 years. It is impossible to predict how many species will disappear when reefs are radically reduced in size. One reason for this is that the vast majority of species are very small invertebrates.

Lower invertebrates are not included in the estimates given by Freeman and Herron (2004), which is not surprising because estimates for such small organisms are practically impossible; not to mention the fact that most species have not even yet been described, and this refers also and particularly to parasites, which probably represent the majority of species on Earth. On pp. 43–46 I have given some estimates for species numbers based on surveys of marine fish. The approximately 14 000 species of marine fish probably have at least 150 000 species of parasites, and the parasite fauna of many other groups of vertebrates is at least as rich. Rohde (2001c) gives data for various groups. The Platyhelminthes (flatworms) alone comprise at least 80 000 species, of which a minimum of 65 000 are parasitic. The number of described species is only a small fraction of these (several thousand), and estimates of extinction rates are impossible because species are small and lack substantial hard skeletons, i.e., they do not fossilize. Estimates of the number of nematode species vary greatly, not only because they are small and many of them are parasitic, but also because a vast number (possibly millions) are thought to live in habitats which have hardly been examined, such as the deep sea and the sand fauna of beaches. The interstitial sand fauna (meiofauna) of coasts, which consists of very small animals, has been examined more or less thoroughly only at one locality, the island of Sylt in the North Sea; 652 species have been recorded to date, and another 200 are thought to exist (Armonies and Reise 2000).

In other words, meiofaunal richness must be enormous, considering the fact that ubiquitous, widespread species seem to be rare.

There are many cases of extinctions which are directly man-induced, for example due to hunting or the introduction of predators, but perhaps more important is habitat destruction. As pointed out by Freeman and Herron, most cases of extinctions in the list compiled by Smith *et al.* (1993), occurred on islands due to the introduction of predators or competitors. However, whereas all but one of bird extinctions in Australia have been on islands, some endangered mammals now occur only on islands (Hugh Ford, personal communication). New introductions are still continuing (e.g., the cane toad, Indian myna, deer in Australia), but they are apparently declining on land, where extinctions due to habitat destruction are now more important. On the other hand, introductions of marine organisms may even be accelerating because of increasing rates of transport by ships and increasing volumes of ballast water that are released by ships: we have discussed the example of the green crab on pp. 46–47. Little is known about how habitat destruction affects invertebrates, because attention has almost entirely focused on large species, and in particular large plants and vertebrates. However, concerning parasites, it is obvious that extinction of a single vertebrate (or invertebrate) species leads to the extinction of a whole parasite fauna, because many of the parasites are strictly host specific. For example, there are well over 100 species of protistan and metazoan parasites of man, many of them strictly specific to man. Of the cestodes infecting man, *Diphyllothothrium latum* is also found in a wide range of fish-eating mammals, but *Taenia saginata* and *T. solium* are only found in humans. Among the numerous nematodes of man, the pinworm *Enterobius vermicularis*, and the large roundworm *Ascaris lumbricoides* only infect man (*A. suis* of pigs is probably a different species). Also, it is not unlikely that even a reduction in population size of hosts may lead to the disappearance of parasites, because a certain population density of hosts may be necessary to safeguard the survival of a parasite species. However, these aspects have been little studied.

All this means that estimates of extinction are with certainty severe underestimates (because host-specific parasite species have never been considered in such estimates), and projections into the future are even more difficult than commonly thought (because the factors leading to parasite extinction, and to extinction of small free-living invertebrates, are poorly understood). Nevertheless, knowledge is sufficient to conclude that we have indeed entered a period of widespread extinctions at least at the level of large plants and animals, and that extinctions are likely to continue particularly because regions of highest diversity in the tropics are strongly affected (e.g., deforestation of the Amazon rainforests, destruction of coral reefs as the result of human activities leading to coral bleaching, etc.) resulting in conditions that resemble those in earlier eras of mass extinctions discussed earlier in this chapter. Also, these conditions were typical of nonsaturation and evolutionary nonequilibrium; it took many millions of years until species numbers recovered and were finally exceeding numbers before the extinctions.

A better-founded view on how widespread extinctions really are could be reached by obtaining quantitative data (even if only approximate) for the various parameters in the diversification equation discussed above. In particular, it is important to note that most parasite groups are much more diverse in tropical than in cold environments, and some (although probably not all) parasite groups are also *relatively* more diverse at low latitudes (i.e., there are more parasite species per host species at low than at high latitudes), as demonstrated by Rohde and Heap (1998) for monogenean gill parasites of marine teleost fish. Whereas there are about as many monogenean species as marine teleost species on the gills at high latitudes, in the tropics there are twice as many. Therefore, mass extinctions in tropical "hot spots" will have a magnified impact on parasite faunas.

In conclusion, we can state that periods of stasis, i.e., apparent equilibrium, existed in evolutionary history, at least for marine benthos which is the best known. There were also repeated severe reductions in diversity (mass extinctions). However, overall, these periods were superimposed on a "walk" to ever increasing diversity which has continued to the Recent. This walk is now threatened by human impacts that have caused mass extinctions over thousands of years at accelerating speed. Mass extinctions are likely to accelerate even further if present trends continue or become stronger (e.g., due to global warming), resulting in severe evolutionary nonequilibrium. Estimates of species lost or threatened are underestimates, because parasites, which constitute a very large proportion of the total fauna, have been largely ignored.

7 · *Some detailed examples at the population/metapopulation level*

In the following chapters, I discuss some examples of equilibrium and nonequilibrium conditions in populations and communities in greater detail. Some studies deal with both populations and communities. Those with emphasis on populations are discussed in this chapter, those with emphasis on communities in Chapter 8.

Reef fishes: density dependence and equilibrium in populations?

Coral reefs are among the most diverse ecological systems on Earth. Species include not only a great range of fishes but numerous invertebrates and plants. For example, Heron Island, at the southern end of the Great Barrier Reef is a small coral cay (island) 900 × 300 m in size, with a reef about 16 km long by a few km wide around it. Nobody has ever counted the number of invertebrate species, most of which probably have not been described, but more than 900 species of fishes have been recorded there. Knowledge of how such diverse systems function is of paramount importance, and, indeed, a very large number of studies on the behavior and ecology of reef fishes in the Indo-Pacific and Atlantic Oceans have been conducted using a variety of methods. Nevertheless, agreement on some important aspects of reef fish ecology has not been reached, largely due to the great range of habitats and species, and partly due to confusion about concepts and theoretical interpretations of the findings. I will discuss some studies to show the often contradictory findings and interpretations.

Many but not all reef fishes live in open systems, producing numerous dispersal larvae with high mortality. Many species are non-territorial, juveniles of others settle and become strictly territorial.

Sale and others, in a number of important papers, investigated reef fish assemblages on the Great Barrier Reef (and later in other regions), using experiments and long-term monitoring. Populations studied were often below carrying capacity, and, according to Sale (1991), on a small scale, i.e., that of a few meters, reef fish assemblages appear to be nonequilibrial showing great variability over time and space. Territorial fish are continuously competing for space, defending their small territories which serve as sources of food, shelter, and nest sites. Territories are dead rock surfaces covered by algae. In one study, three species of pomacentrid fishes were studied. They inhabited territories that were contiguous but did not overlap. Over 40 months, the total habitat area remained more or less the same. Pelagic juveniles or older juveniles from other sites occupied territories that had become vacant by mortality. However, the species occupying particular territories changed, indicating that recruitment was by chance or "lottery" (who comes first, wins). Species involved in such lotteries must be iteroparous. Sale referred to such lotteries as "non-equilibrial," other authors as "equilibrial." In another study, Sale *et al.* (1984) made a 2-year visual census of young of the year recruits to 2 habitats each, on 7 reefs. Selected sites were very similar. Nevertheless, there was significant variation in abundance between reefs, between years, or both. Recruitment variability was also documented in several subsequent studies reviewed by Doherty and Williams (1988). Sale *et al.* (1984) reported results of a 9-year study of assemblage structure of fish from 20 coral patch reefs, based on 20 non-manipulative censuses. The total number of fish species found was 141, belonging to 34 families. The average reef was about 8.5 m^2 in surface area and had about 125 fish of 20 species. Variation in total fish numbers among censuses was at least 2-fold, but reached 10-fold in 12 reefs. Variation in composition and relative abundance of species was also great. This variability could not be attributed to changing physical structure of the reefs, or to the effect of rare species leaving or entering the reefs.

Sale and Tolimieri (2000) questioned the necessity of density dependence for long-term persistence in reef fish populations. Like Sale above, they emphasize the nonequilibrium aspects of reef fish populations. There are situations in which demographic rates change in relation to density without negative feedback. Thus, per capita recruitment will decrease with increasing population size although absolute numbers of recruited larvae do not change. Furthermore, there are many density-independent processes, and many existing populations do not persist. Also, populations do not automatically bounce back from a low population size.

Several recent studies appear to support the view of great variability in reef fish assemblages. Chittaro and Sale (2003) compared reef fish assemblages at small patch reefs in the Caribbean (St. Croix, US Virgin Islands) and on the Great Barrier Reef (One Tree Reef) and found very substantial changes in assemblage structure over time in both localities, despite the differences in taxonomic composition. Conclusions were based on monitoring the fish populations over 10 years at One Tree Reef and over 5 years at St. Croix. A recent study of early recruitment of coral reef fish in the tropical eastern Pacific, at a locality (Gorgona Island) where tidal amplitude reaches 4.4 m, has shown that lunar and tidal factors contribute to recruitment dynamics. In species with sporadic and aperiodic recruitment pulses, stochasticity in weather conditions and in oceanographic events may be the most important factor determining recruitment variability (Lozano and Zapata 2003).

Robertson and co-workers made thorough studies of reef fish assemblages and arrived at conclusions somewhat different from those of Sale. For example, Robertson (1995) doubted the existence of a lottery mechanism at least for the Caribbean damselfish *Stegastes diancaeus* and *S. dorsopunicans*. The former species is larger, grows faster, and lives longer than the latter, with which it shares its primary habitat, and from smaller individuals of which it often takes over habitat. There was no evidence that the competitive advantage due to larger body size was offset by other competitive advantages. Asymmetric competition between other closely related species of the same genus with similar ecological requirements was also shown to limit abundances of species (Robertson 1996). Robertson and Gaines (1986) demonstrated the important role of interspecific competition in surgeon fish at the outer edge of the barrier reef of Aldabra, Indian Ocean. When meeting, 38 pairs interacted rarely and 27 pairs always interacted. With the exception of one pair, all pairs had highly asymmetric dominance relations. Subordinate species took over habitats vacated by previously agonistically dominant species, but the reverse was rare. The authors concluded that interference competition for food plays a role in structuring reef fish assemblages: 60–80 % of habitat use relationships in the area of study may be affected. Competition may also occur between taxonomically totally different species. In 1983/84, 95 % of the sea urchin *Diadema antillarum* in Atlantic Panama died off, which led to a sharp increase in soft corals, some of which are also eaten by reef fish. Three herbivorous fish species, *Acanthurus* spp., were monitored between 1978 and 1990 on six patch reefs. The two fish species that feed largely on reef substrata, increased by 250 and 160 %, respectively. The third species, which often feeds off the reefs, was not affected (Robertson 1991).

Robertson (2001) demonstrated that taxonomically and biologically, endemic reef fish from seven small islands are broadly representative of the regional fauna, suggesting that many reef fishes could have persistent self-sustaining local populations.

All studies of reef fish have been conducted at the level of local populations. Sources of density dependence may be fecundity, mortality of eggs and larvae before settlement, mortality during transition from pelagic to demersal stage, or mortality of juveniles/adults after settlement (Hixon and Webster 2002). According to these authors, various studies have confirmed that growth of reef fish and hence fecundity are density dependent. Mortality during settlement can be considerable, and recruitment to local populations is per capita density dependent, i.e., a decreasing proportion of the total population is recruited with increasing size of the local population. On the other hand, rates of loss of eggs and larvae during presettlement mortality seem to be density-independent but are little known. Therefore, according to the authors, density dependence must act at a later stage in order to guarantee persistence of the metapopulation. Mortality at postsettlement is better documented, although few studies deal with mortality due to interspecific competition. Predation, on the other hand, has been shown to cause density-dependent mortality, which supports the recruitment limitation hypothesis. The authors surveyed the literature and found a broad range of results concerning recruitment patterns both in observational and experimental studies. Recruitment in reef fishes appears to be highly variable, but after conversion to per capita rates, recruitment in all cases seemed to be "pseudo-density dependent." A survey of postsettlement mortality permitted calculation of the instantaneous per capita mortality rate as a function of local population density. Fourteen observational and experimental studies from 24 separate data sets on 15 fish species were evaluated. The evaluation showed that postsettlement density-dependent mortality is indeed widespread among reef fishes. Combined consideration of recruitment and mortality led to the conclusion that local population fluctuations are bounded as the result of density-dependent processes. Figure 7.1 shows this combined effect for two species of reef fish, suggesting regulation of local populations. Equilibrium is reached where recruitment and mortality lines cross, provided there is no emigration or immigration. Concerning the mechanisms involved in density-dependent regulation, data for recruitment are lacking but may be intraspecific competition and/or cannibalism. For predation, the most important source of mortality in reef fishes, density-dependent effects have been demonstrated in several studies. For example, predators aggregate at sites where new recruits are common, whereas sites with few recruits are ignored.

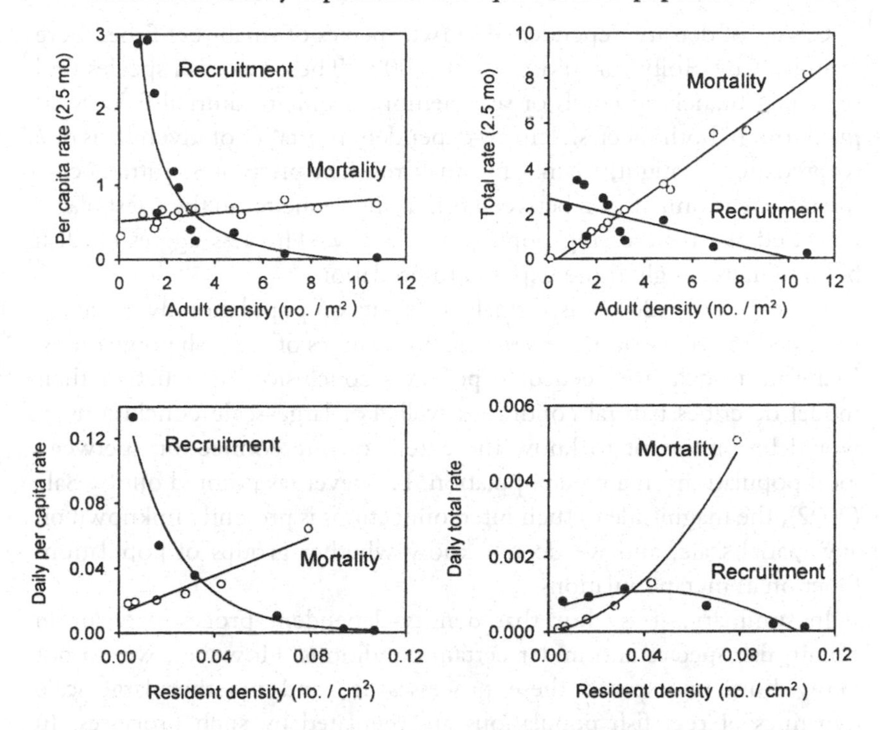

Figure 7.1. Comparison of recruitment and mortality curves presented as per capita rates (left) and as total rates (right) of two fish species in the British Virgin Islands. Equilibrium is reached where recruitment and mortality lines cross. Data from Forrester and Schmitt and Holbrook, from Hixon and Webster (2002) in Sale, P. F. (ed.) (2002). Reprinted by permission of the authors and Elsevier.

Doherty (2002) found a curvilinear relationship in many studies and for several species, consistent with an asymptotic limit to abundance. Density dependence is also shown in cohorts of newly settled fish, and the author – following Murdoch (1994) – assumes that fish stocks must experience density-dependent "regulation" of some scale to prevent unbounded random walks to extinction or overpopulation, although local populations may also be recruitment-limited. He cites Armsworth (in press), according to whom density dependence is not required of any demographic sink that depends on external inputs for its persistence. However, primary recruitment limitation is only possible when there is low settlement relative to potential saturation for an additional cohort. If this is the case, abundance is simply the product of settlement and density-independent mortality (Werner and Hughes 1988). Armsworth (2002, cit. Doherty) presented mathematical proof that metapopulations require density dependence on some but not all scales.

Causes of density dependence in two species of small reef fishes were examined by Holbrook and Schmitt (2002). These two fish species find shelter in branching corals or sea anemones, and are diurnal feeders on plankton. In both species, density-dependent mortality of juveniles is due to predation at night, mainly by small resident piscivores. Intraspecific interference competition between fish trying to shelter in the safest places increased with increasing population density. The less aggressive fish became increasingly more exposed to predation.

Forrester *et al.* (2002) used small-scale data on population dynamics, as discussed above, to simulate large-scale dynamics of reef fish populations. Many more data are needed to permit a conclusion on whether their model describes natural conditions well. For large-scale conclusions, it would be important to know the extent of interconnections between local populations in a metapopulation. However, as pointed out by Sale (2002), the magnitude of such interconnections is presently unknown on any spatial scale, and we do not know whether groups of populations function as metapopulations.

In summary, it is clear that density-dependent processes occur in certain fish species and under certain conditions. However, we do not know how widespread these processes are and whether large-scale dynamics of reef fish populations are regulated by such processes. In particular, the great variability between regions makes generalizations difficult. Considering these difficulties, agreement on contentious issues may be impossible in the foreseeable future.

Kangaroos: fluctuations in rainfall are the primary determinant of population size, but there is some "regulation" by negative feedback

Several species of kangaroo are found in Australia. This discussion is restricted to the red kangaroo, *Macropus rufus*. The species is widely distributed throughout inland Australia, but its greatest density is reached in the eastern sheep rangelands, for example in northwestern New South Wales where it can exceed 20/km^2. The total number for Australia in 1981 was estimated to be 8.4 million (Caughley 1987a). Because it competes with sheep (total number in 1981, 133.3 million), it has been studied extensively, and much of the information is reviewed in Caughley *et al.* (1987). Many of the study sites are sheep pastures, where kangaroos and sheep live together, others are in National Parks, which lack sheep. Male red kangaroos reach a weight of about 83 kg, females 39 kg (Stuart

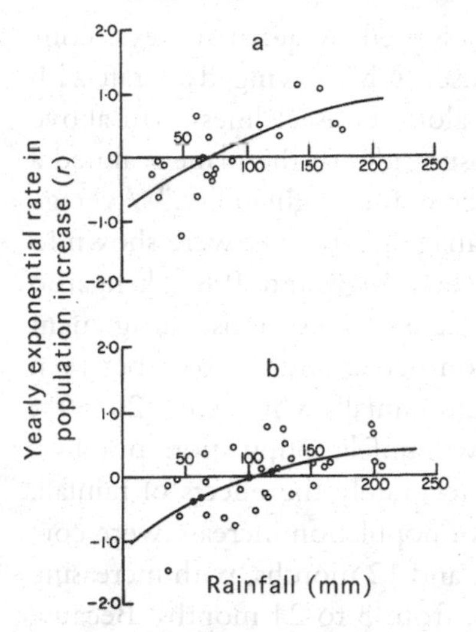

Figure 7.2. Numerical responses, i.e., changes in annual rates of population growth, (Mitscherlich equations) of red kangaroo populations (1978–1984) to summer–autumn rainfall in (a) the Western kangaroo management zone of South Australia, and (b) the Central and Eastern kangaroo management zones of South Australia. In both examples, annual rates of population growth increase with rainfall, but at very high rainfalls the increase becomes more and more asymptotic. From Cairns and Grigg (1993). Reprinted by permission of the authors and Taylor & Francis.

Cairns, personal communication). Although some individuals move over large distances (maximum recorded, over 300 km), most appear to remain in a relatively small area of a few (<8) km^2 (Priddle 1987, references therein). Reproduction appears to be more or less continuous, the maximum number of young per adult female is about 1.5 per annum. The diet of *Macropus rufus* consists of grasses, forbs, chenopod sub-shrubs, and some shrubs (Barker 1987). Potential competitors for food are sheep and rabbits. Grazing efficiency of red kangaroos, sheep, and rabbits is about the same, they all reduce biomass to about 20–50 kg/ha, depending on the biomass of unpalatable plant species present. They also need about the same amount of food per day, when expressed as metabolic weight (60–80 g/$kg^{0.75}$/day), and when food is not limited. At biomasses of above 300 kg/ha, competition between these herbivores for food is minimal (Short 1987). Availability of food depends on rainfall; Figure 7.2 shows examples of numerical responses (changes in annual rates of population increase, *r*) of red kangaroo populations to variations in summer–autumn rainfall in two regions in the South Australian pastoral zone, surveyed from 1978 to 1984. Populations were assessed by aerial surveys, conducted within 3–4 hours of sunrise or sunset. A high-winged aircraft with a speed of 185 km/h was used. It flew along transect lines 76 m above ground. Two observers sitting on opposite sides of the plane scanned a 200 m wide strip. Counts were corrected for "sightability." Average population densities in the kangaroo management zones were shown to range from 0.2 km^2 to 11.3 km^2. Between 1978 and 1981, kangaroo numbers increased, although rates of increase in the various management zones differed. In some zones, population size almost doubled between 1980 and 1981, coincident with very high rainfalls which were 20–50 % above average. During periods of low rainfall, population numbers declined. In order to determine more accurately the effects of rainfall, the calculated annual exponential rates of population increase were correlated with rainfall for intervals of 3, 6, and 12 months, with increasing time-lags ranging, in steps of 3 months, from 3 to 24 months. Because there must be a limit to the rate of increase, the form of the numerical response was expected to be asymptotic. Hence, a model representing such asymptotic relationships, i.e., the Mitscherlich equation was selected (for details see Cairns and Grigg 1993, further references therein). Figure 7.2 shows that increases of reproductive rates level off, i.e., become asymptotic, when rainfall becomes very high. The most consistent and strongest positive correlations between rates of population increase and rainfall were obtained for the shortest time-lag. Cairns and Grigg (1993) discuss the possible mechanisms causing population fluctuations. They consider it likely that broad-scale shortage of food in a drought may directly lead to the death of more vulnerable individuals. There may also be effects on effective natality: females may enter anoestrus, and pouch young may not survive. Generally, drought may result in heavy juvenile mortality and, to a lesser extent, heavy mortality among the very old kangaroos. Also, more males than females die (as shown in other surveys).

Rainfall is the predominant factor determining population densities of the red kangaroo, as shown in Figure 7.3, which illustrates a 100-year run for a system for which certain assumptions concerning the numerical responses of kangaroos to pasture biomass (which in turn is determined by annual rainfall) have been made (Caughley 1987b). Different assumptions may modify the exact shape of the curve representing kangaroo densities, but the general shape remains the same. It is important to note that the annual changes in pasture biomass in this model are a product of interactions between rainfall over the year and herbivore density. Herbivore

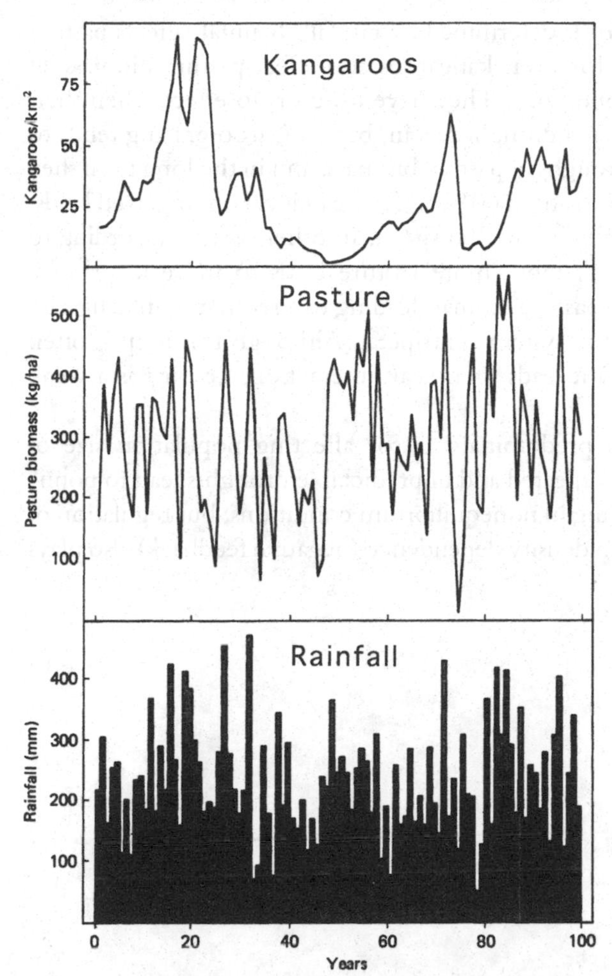

Figure 7.3. Predictions for the relationship between annual rainfall (bottom), pasture biomass (middle) and kangaroo densities (top) over a 100-year period, using a model developed on the basis of long-term empirical studies. Slight changes in the assumptions do not lead to fundamentally different predictions. From Caughley (1987b). In: Caughley, Shepherd, and Short, eds., 1987. Reprinted by permission of Cambridge University Press.

density is determined not by present conditions but by previous biomass, which again is the result of earlier rainfall.

Caughley (1987b) discusses the question of how kangaroo populations are "regulated." At first glance, it seems that populations are not regulated, because of the irregular fluctuations in density, dependent on pasture biomass, which in turn is determined by rainfall. Rainfall affects pasture with little time-lag. However, kangaroos also affect pasture biomass, at least under certain conditions. They have little or no effect when their density is low, or when a drought sets in, but kangaroo grazing tends to lower the peaks and troughs of pasture biomass, and in the long term they lower pasture biomass to about 60 % of the mean level on ungrazed lands, from about 500 kg/ha to about 300 kg/ha. In other words, according to Caughley, the feedback loop (more pasture leads to more kangaroos, which in turn reduces pasture biomass leading to a reduction in kangaroo density, etc.), makes the system centripetal. Although this loop is often weak, in the long run it leads to a weak interaction, i.e., it has a weak regulatory effect.

In conclusion, the predominant factor affecting population size of kangaroos is rainfall. Repeated and unpredictable droughts lead to population collapses, resulting in nonequilibrium conditions; but regulation of population sizes due to density dependence (negative feedback) also plays some role.

8 · *Some detailed examples at the community level*

Tropical rainforests: how is diversity maintained?

General considerations

Tropical rainforests are among the most diverse ecosystems on land. Amazonian rainforests, for example, can have more than 280 species of trees in one hectare (with a diameter at breast height of at least 10 cm), and a 0.52 km^2 plot in Borneo had 1175 species (with a diameter at breast height of at least 1 cm) (references in Wright 2002). Diversity further down in the hierarchy is considerably greater, and it is much more difficult to assess: many if not most species have not yet even been described. For these reasons, most detailed long-term ecological studies that have been conducted in rainforests are concerned with the larger species, i.e., trees. Wright (2002) reviews much of the work done on mechanisms that permit coexistence of so many species in tropical rainforests. Connell and various co-workers have made long-term studies of rainforests and equally diverse marine systems, i.e., coral reefs in eastern Australia, extending over more than 30 years. I discuss their work on tropical and subtropical rainforests in northern and southeastern Queensland in some detail, because it tests clearly formulated hypotheses using long-term extensive and intensive data sets from habitats little disturbed by man (Connell 1978, 1979; Connell *et al.* 1984; Connell and Green 2000).

The main problem Connell addresses is the degree to which equilibrium or nonequilibrium conditions contribute to the great diversity in tropical rainforests. Nonequilibrium may, for example, be due to frequent and reasonably strong disturbances that prevent the establishment of equilibrium, and thus lead to the survival of many competitively inferior species. Equilibrium will develop when such disruptions are absent, or weak and infrequent, and when compensatory mechanisms that prevent replacement of rare species by one or a few dominant ones are not effective. A further problem addressed is the significance of spatio-temporal heterogeneity in maintaining diversity.

It is important to understand details of the line of argument used by Connell. Connell (1979) discusses the theoretical background for his studies. Following Caswell (1978), he distinguishes between closed and open systems. In a closed system, a population is found in a single, fairly homogeneous habitat without immigration. In an open system, populations in different "cells" are linked by migration of individuals between them, although populations are still more or less independent, since exchange of individuals is relatively limited. This corresponds to what is usually defined as a metapopulation, a term not used by Connell. The "open" system seen as a whole may well be a closed system, in spite of the open nature of its components. Local extinction in open but not in closed systems can be overcome by recolonization.

The definition of equilibrium is scale-dependent. Connell follows the usage of Caswell, according to whom equilibrium applies to what is happening within a habitat cell: a system is equilibrial when, for example, competitors within a cell can coexist as a result of some underlying mechanism. However, a whole system can also be considered to be equilibrial, even when one competitor always out-competes another in a particular cell, as long as competitors coexist indefinitely in the system.

Connell defines "gaps" as empty cells created by disturbances. A gap will first be invaded by species that are adapted to early invasion. Once a gap is filled, these early colonizers cannot successfully immigrate any longer, because their adaptations to early immigration (such as their ability to germinate or attach in exposed habitats, and high metabolic rates) are not suited to late invasion under changed conditions. However, if, for example, major disturbances are repeated, early invaders will have the chance to colonize newly created gaps and the whole system remains open and in nonequilibrium. In contrast, lack of further major disturbances will lead to the dying out of early colonizers, i.e., for these species, the habitat is now a closed nonequilibrium one. Repeated minor disturbances will permit intermediate and late colonizers to fill the gaps, i.e., for them the system is open and nonequilibrial. If disturbances are very rare, only late-succession species can survive, and the system is either open nonequilibrial or closed equilibrial. Nonequilibrium in this case depends on the existence of "compensatory mechanisms" that permit coexistence of species. If such mechanisms do not operate, one species will eliminate all others, leading to a closed and equilibrial system. Connell mentions as a possible compensatory mechanism frequency-dependent mortality due to an increase in predation at high prey population densities. One could add frequency-dependent mortality due to greater risk of disease caused by parasites or microorganisms. Connell mentions another mechanism, that of "circular networks." A circular network exists, for example, if species 1 out-competes species 2, species 2 out-competes species 3, but species 3 out-competes species 1. A different competitive mechanism must be involved for at least one species in the chain. Co-occurrence in a single cell is not possible, because one species will finally win unless the various species are exactly balanced in their chances to survive, a highly unlikely event. If species occur in pairs in different cells in an open, locally nonequilibrial system, the whole system will stay in equilibrium if dispersal and extinction rates are the same for the species, even though in single cells only one species will survive. In other words, these compensatory mechanisms work only in open systems.

Connell (1979) summarizes these considerations as follows. Early colonizers (which cannot invade occupied sites) are abundant and dominate the cover if moderate to large disturbances occur frequently; they are not replaced and die out when disturbances are small and frequent, or when infrequent disturbances occur at an intermediate time after a large disturbance; early colonizers are absent when disturbances are very small and infrequent, or when disturbances occur a long time after a large disturbance. Late colonizers (which cannot invade open, exposed sites) are absent or rare (near gap edges) when disturbances are moderate to large and frequent; they are common and invade in gaps or shade, when disturbances are small and frequent, or when infrequent disturbances occur at an intermediate time after a large disturbance; late colonizers occur in open nonequilibrium or open equilibrium if compensatory mechanisms permit coexistence, or they occur in closed equilibrium with a single dominant species, if compensatory mechanisms are not operative.

Connell (1979) then evaluates evidence from tropical rainforests, based on his own studies and those of others, to explain how these theoretical considerations conform to what is happening in nature. Firstly, he points out – referring to work by Eggeling on a tropical rainforest in Uganda, and by Jones on a tropical rainforest in Nigeria – that evidence supports the view that a single species will become dominant if disturbances are rare: the canopy in both rainforests was dominated by large, old trees of a few species. In the Uganda rainforest, colonizing, mixed, and climax stands could be distinguished. In the colonizing stand spreading into adjacent grassland, the canopy was dominated by a few species, juvenile trees belonging to different species. Tropical rainforests generally contain some species that have many large trees and few offspring. It is likely that they will be replaced by different species having more offspring. Connell interprets this as meaning that high-diversity tropical rainforests are in a nonequilibrium intermediate stage after a disturbance. Experimental evidence supports this view. Connell describes an experiment by Webb, Tracey, and Williams in a Queensland rainforest. In an experimental clearing, over a 12-year period, a uniform pattern of a few early colonists changed to a patchy pattern of many species of somewhat later colonists. At least some tropical forests have established a closed equilibrium. Thus, in the Uganda rainforest discussed earlier, the canopy of climax stands was dominated by 75–90% ironwood, with much ironwood offspring also in the understorey, suggesting closed equilibrium with a single dominant species.

Compensatory mechanisms

Concerning the compensatory mechanisms permitting coexistence of species in the absence of environmental disruptions, Connell (1979) evaluates the evidence for frequency-dependent mortality, which could be responsible for coexistence of many species, and which could, for instance, be due to greater mortality of seedlings occurring in dense clumps or near conspecific adult trees. He found no unequivocal evidence for density-dependent mortality. Mortality was either greater in high-density patches, or it was equal. Concerning the effects of density of young trees on mortality, Connell examined 49 tree species from subtropical South Queensland and 46 tree species from tropical North Queensland. The mortality rate was determined by plotting the mortality rate between 1965 and 1974 against the original number mapped in 1965. No correlation was found. Also, according to Hubbell *et al.* (1990) density-dependent effects could not be important in determining diversity since most tropical species are rare. Furthermore, Hubbell (1980) and others demonstrated that differences in per capita recruitment due to distance or density dependence are minor in comparison with the effects of larger numbers of seeds falling near trees: experiments have shown that recruitment was greatest where seed density was greatest.

Concerning the effects of near conspecifics, Connell showed that seedlings and saplings having the same species as nearest neighbors had greater mortality than those mixed with other species. Nevertheless,

Connell (1979) concludes that, since some species are and others are not controlled by this mechanism, species with less or equal mortality near the parent will have a competitive advantage and displace the others, i.e., this mechanism cannot explain the great diversity.

Further support for the conclusion that density effects cannot give an explanation of great diversity comes from a very detailed study of Connell and Green (2000), who examined the dynamics of seedlings of a population of the tropical rainforest tree *Chrysophyllum* sp. nov. (Sapotaceae) from tropical Queensland, using data for a 32-year period. The tree is a common, shade-tolerant canopy species. The study plot and its surroundings had never been disturbed by humans to any important degree. The spatial scales considered ranged from 1 m^2 to 1 ha, and the temporal scales ranged from 1 to 32 years. Of the new recruits, 98.3% were found in six so-called "mast" episodes; in the nine censuses conducted in other years only 1.7% were found. In mast years, seed fall and recruitment of seedlings were often dense and patchy, with a maximum of newly germinated seedlings of 140/m^2 , and more than 25/m^2 in areas covering tens of square meters. Dense patches of recruits were found in different spots at each succeeding mast episode. Analysis of the data showed that initial seed density had no effect on survival to germination, and growth decreased with increasing density in one year, but not in another. Concerning greater mortality near conspecifics, no significant trend in mortality with increasing distance from conspecific adult trees was found; however, when all other species were combined in the analysis, mortality of seedlings decreased with increasing distance from *Chrysophyllum* trees.

Connell *et al.* (1984) found clear evidence for compensatory mechanisms, at least for some age groups of trees. The authors examined the possible contribution of compensatory mechanisms towards maintaining high diversity in two rainforest plots in eastern Australia, one in tropical North Queensland (elevation about 850 m), the other in the subtropical Lamington National Park, southeastern Queensland (elevation about 900 m), beginning in 1963. Results for both areas were very similar. In each area, trees were mapped along parallel survey lines 20.1 m apart, and various methods, not discussed here, were used to analyze tree diversity, abundance, etc. of various size classes, to answer the general question whether rarer species are favoured over more common ones by some kinds of compensatory mechanisms. More specifically, two hypotheses were tested: according to the first hypothesis, more common species have lower rates of recruitment and growth, and greater mortality than rarer species. According to the second hypothesis, trees grow more slowly and die more easily when they are close to a conspecific tree than when they are close to trees of other species.

To test for the first hypothesis, the growth of seedlings of a species was compared with abundance of conspecific adults. Analysis showed that, on the scale of the whole plot, abundance of adult trees had little effect. This hypothesis was further tested by comparing growth in each species with abundance and average size within the same size class. The results showed that growth was never correlated with abundance. The hypothesis was also tested by relating abundance of adults with mortality of conspecific seedlings. No correlation was found. Furthermore, 44 out of 48 tests for distance to first and second nearest neighbor did not show an effect. If there was an effect, its direction was not consistent. So, no further tests for the effect of species of neighbor were made.

To test for the second hypothesis, growth rates of individuals with conspecific neighbors were compared with growth rates of individuals with neighbors of different species. In one, 11 out of 12 comparisons and in the other 8 out of 12 comparisons were negative. However, a distinct compensatory effect was found in the four comparisons of the two most recent seedling classes in the tropical (northern Queensland) rainforest area: growth was significantly faster when the first or second nearest neighbor was a different species. Connell *et al.* concluded that a compensatory mechanism may operate but only at a very early stage of growth. In 13 out of 20 comparisons of seedlings and saplings in the tropical rainforest, and in 12 out of 24 comparisons in the subtropical one, mortality was greater when the nearest neighbor was a conspecific. Mortality was never greater when the nearest neighbor belonged to a different species. The second hypothesis was further tested by comparing the effects of increasing density of common species on growth and mortality of seedlings and saplings. The results showed practically no (or at best a very weak) effect on growth and mortality. The authors also conducted field experiments to test the hypothesis, using two plant species. Seeds were placed on the ground either densely or sparsely. No significant differences were found between the two treatments, although almost all plants disappeared over the observation period of 9.5 months. The second hypothesis was also tested at the scale of individual trees by calculating mortality of seedlings and saplings in the inner circle, the outer ring around each conspecific adult tree, and in the area beyond. The inner circle is the area within 1.5 times the crown radius of the adult, the outer ring is an annulus concentric to and of width equal to the radius of the inner circle. In only 1 out of 23 comparisons was mortality greater in the inner circle than the outer ring. Only 5 out of 100 comparisons between the two inner zones and the area "beyond" showed greater mortality, and twice mortality was even lower. The authors conclude that the few cases of compensation could perhaps be explained by chance. Field experiments were conducted using seedlings of two tree species. In one, mortality was closer to a conspecific adult, in the other there was no difference.

The authors also tested for intermingling of species. As indicated above, seedlings of common species have greater mortality near conspecific adults. This could result in creating opportunities for rarer species and lead, in time, to a higher degree of intermingling. Use of Pielou's test for "pattern diversity of communities" showed that individuals of all sizes have conspecifics as neighbors more often than expected by chance for most size classes of seedlings. However, pattern diversity increased significantly with age, and it also increased significantly in the period since initial mapping, although among the eight older classes, there was only one significant increase. The conclusion has to be that most intermingling occurs between the seedling and sapling stages.

Another possible compensatory mechanism (later called the Janzen–Connell effect by others) is that herbivores could be attracted by high-density clumps composed of single species or by seedlings close to adults of the same species. Connell (1979) dismisses the effectiveness of this mechanism, because it implies a great degree of specialization: herbivores must be able to distinguish between tree species for it to become effective. Herbivores in tropical forests were indeed shown to have low host specificity. Data from 900 herbivore species feeding on 51 plant species in New Guinea were analyzed by Novotny *et al.* (2002), and most were found on several plant species. According to the authors, monophagous herbivores are probably rare in tropical forests, because species-rich genera are dominant. Also, most herbivore communities shared a third of their species between phylogenetically distant hosts. However, differential preferences of herbivores for different host species were not considered. (It can be done by using host specificity indices, Rohde 1980e; Rohde and Rohde 2005.) On the other hand, experiments showed poorer performance of seeds or seedlings near conspecific adults in 15 out of 19 populations with an insect as the main herbivore; it was worse in 2 out of 27 populations with a vertebrate as the main herbivore (references in Wright 2002). Nevertheless, Hubbell (1980) argued that the Janzen–Connell effect could not be important, considering its short-distance effect and the great species diversity, an argument not accepted by all (see below, Wright).

A further possible compensatory mechanism mentioned above is that of circular networks (which involves at least two different competitive mechanisms between species). Connell points out that rainforest trees in the tropics are too similar for such a mechanism to be likely, although it remains an as yet unproven possibility.

Connell *et al.* (1984) concluded from the evidence available at the time that compensatory mechanisms seem to be operative at some spatial scales, but not at others. Over the whole area, seedling recruitment of subcanopy and understorey, but not of canopy species was indeed higher for rarer species. At the scale between nearest neighbors, there was compensatory growth soon after establishment of seedlings which stopped after about 6 years. Growth was slower close to conspecifics. There was compensatory mortality for most size classes. All of this may prevent local displacement of rarer species by the more common ones. Concerning the mechanisms responsible, the authors discuss intra- and interspecific competition, and indirect effects due to microorganisms and other natural enemies, but a decision as to which mechanisms are responsible is impossible. Concerning the two hypotheses that were tested, these apply to some species but not to others.

Gap dynamics (intermediate disturbance hypothesis)

In spite of evidence for the existence of compensatory mechanisms, Connell (1979) concluded that the most likely explanation for the high diversity in tropical rainforests is the occurrence of frequent small disturbances keeping the system open and in local nonequilibrium. Regional equilibrium may (or may not) be brought about permitting the persistence of less shade-tolerant species among the more tolerant ones. Greatest diversity can be expected in open-nonequilibrium systems that are at an intermediate stage after a severe disturbance, or which are exposed to smaller and neither very frequent nor very infrequent disturbances. Connell names this the "intermediate disturbance hypothesis." However, as stressed by Connell *et al.* (1984), there is no way at this stage to determine the relative importance of compensatory mechanisms, environmental heterogeneity, intermediate disturbances, gradual climatic change, random variation in conditions affecting reproductive and mortality rates, among others.

Moreover, Wright (2002) has pointed out that very few of the vast number of species of tropical rainforest trees have been examined with regard to gap dynamics. Criticism of the hypothesis is based on experimental evidence and theoretical considerations. Since adults of species with different regeneration requirements often occur close together, application of the hypothesis may be limited. Reasons for the closeness of such species may be that environmental gradients close to each other in gaps may be quite different, and that new gaps may be superimposed on already existing ones. Experimental and observational evidence given by Wright against the importance of the hypothesis are that, in one experiment, more than 80 % of shade-tolerant saplings survived after a gap had been formed, and that regeneration of most trees is so slow that they do not reach the canopy in a single gap cycle.

Environmental heterogeneity

The discussion so far has ignored the effect of environmental heterogeneity. Species coexistence might be possible in heterogeneous habitats even in the absence of compensatory mechanisms. Theoretically, each species in a rich assemblage of species may be adapted to a different component of the heterogeneous habitat. For example, one species might prefer a slightly more acid soil than the others, another species might prefer a harder over a softer substrate, etc. But heterogeneity may not only be spatial, i.e., exist at a local site, it may also be temporal, due to climate change. The crucial question is: how important is heterogeneity in determining diversity relative to environmental disturbances? Connell (1979) argues that it is impossible to test the hypothesis. Among the difficulties are the practical impossibility of determining the degree of specialization required, and the impossibility of being certain about the niche axes relevant for an evaluation. Furthermore and importantly, it is highly unlikely that the 100 tree species coexisting on 1 hectare of tropical rainforest have niches narrow enough to permit coexistence, in particular since tropical rainforest trees of many species often live intermingled in small areas. Nevertheless, environmental heterogeneity will permit coexistence of some species without the requirement for any additional mechanism. Concerning spatial and temporal variations in the environment, Wright (2002) emphasizes that, among biotic resources, it is unlikely that pollinators and seed dispersal agents are important in promoting plant diversity, because most of these agents are generalized, and very few, such as wasps pollinating figs, are strictly host-specific (see above, Novotny *et al.* 2002). In contrast, spatial heterogeneity is marked and the distribution of many plant species responds to micro-topographic gradients. However, since many plant species are distributed across such gradients, it is highly unlikely that the very great diversity can be explained by spatial heterogeneity alone.

So, the general conclusion from the discussion at this time is that several compensatory mechanisms, as well as environmental heterogeneity, may play some role, as may disturbance at intermediate levels. "Gaps" are commonly produced in tropical rainforests by lightning, strong winds, and landslips, and these gaps are invaded by early colonizers followed by other species, thus maintaining high-diversity communities, since the potentially dominant species never have the chance to completely displace all or most of the other, less dominant species.

Coevolution

Concerning evolutionary factors potentially responsible for the great diversity, Connell (1980) discussed the role of coevolution in rainforest trees and concluded that it is highly unlikely to have any great importance, because there are too many species of nearest neighbors. In a Queensland rainforest, the mean number of tree species with a height greater than 0.5 m per 10 × 10 m plot had 57 other trees among the nearest neighbors. Among the nearest neighbors of less common species, not a single species appeared more often than once. Hence, coevolution between any of these tree species is extremely unlikely.

Some recent work

Wright (2002) reviews much of the work done on mechanisms that permits coexistence of so many species in tropical rainforests, including much work done after Connell's studies were published. He discusses the following, not mutually exclusive, main hypotheses:

(1) Rare species are favoured demographically by compensatory mortality, caused by allelopathy, intraspecific competition, and/or reduction of recruitment near conspecific adults by host-specific pests (the latter referred to as "Janzen–Connell hypothesis");
(2) Species do not have the opportunity to compete because of recruitment limitation, and/or because of low understorey densities;
(3) The environment is neither temporally nor spatially constant, leading to regeneration niches or gaps;
(4) Time for competitive exclusion is insufficient because of intermediate disturbances (Connell) or chance population fluctuations (Hubbell), or because communities are in dynamic equilibrium (Huston);

(5) Growth is not limited by one resource, i.e., different species use different resource ratios;
(6) There is immigration.

Most of the hypotheses were already included in the discussion of Connell's work. I discuss some of the hypotheses listed above that were not included in that discussion.

(1) *Low population density precludes competition*: Wright discusses two experiments testing the hypothesis that low population densities may preclude competition between understorey plants (hypothesis 2 above). In both experiments, some plants were removed and recovery monitored. In neither case was there competitive release in the remaining plants. However, more experiments are needed to confirm these results. Furthermore, also testing hypothesis 2, herbivores do indeed reduce understorey plant density. In the absence of almost all mammalian herbivores, plant density was 230 % greater than in a forest with all of the mammals. Recruitment limitation (hypothesis 2), a hypothesis in particular promoted by Hubbell, is controversial. It may well limit competition between rare species, but is unlikely for common species.
(2) *Dynamic equilibrium model of Huston*: Huston (cited by Wright 2002) suggested that slow population dynamics may permit coexistence of similar species. His dynamic equilibrium model assumes a balance between disturbances and population dynamics. But such a balance has not been demonstrated and is unlikely, because disturbances usually do not reset population sizes, and species diversity is often greatest where population dynamics are the fastest and where resetting of population sizes by disturbances is most unlikely.
(3) *Community drift model and slow competitive displacement*: important theoretical considerations based on experimental evidence are presented by Hubbell and collaborators (Hubbell *et al.* 1990). Hubbell's community drift model was proposed to explain that identical species can coexist indefinitely, assuming that births and deaths occur by chance. However, species are not identical, they do not have the same probabilities of death and reproduction, and it has been shown that the model cannot explain a long coexistence of species (references in Wright 2002). Hubbell and Foster (1986), based on circumstantial evidence, concluded that there is slow competitive displacement in tropical rainforests. Equilibrium hypotheses may explain some of the results obtained, but overall pairwise and predictable interactions between species are not important. More important are disturbances and the effects of biotic uncertainty.

Wright (2002), in his review, concludes that of all the hypotheses listed above, three are strongly supported. First, there is strong support for the importance of niche differences: plants are not randomly distributed along micro-topographical gradients, and there is a tradeoff between survivorship and growth during regeneration. Second, host-specific pests reduce recruitment near conspecific adults (Janzen–Connell effect). Third, negative density dependence over large spatial scales may regulate populations of abundant species. Finally, Wright suggests that a fourth hypothesis may be important, i.e., that competition is reduced because of suppressed understorey plants.

The hypotheses considered to be the best supported do not include gap formation due to disturbances (the intermediate disturbance hypothesis of Connell), although Wright does not rule out its limited applicability. He also does not include the possibility that gradual climatic change may be important. Indeed, such a contribution will be difficult to prove since effects can only be expected over long periods. Even for the hypotheses considered to be the most important, Wright concedes that "implications for species coexistence and plant diversity remain conjectural. Large size, low population density, and long generation times may well preclude experimental evaluation of mechanisms of plant species coexistence in tropical forests."

Nonequilibrium in small animals and plants?

This discussion has been concerned entirely with large trees, for the very simple reason that these are the ones which have been most thoroughly studied because of their large size, and because they lend themselves relatively easily to manipulative experiments. However, the vast majority of tropical rainforest species are much smaller: fungi, algae, microorganisms of various groups among the plants, and a vast array of invertebrates among the animals, most of them not yet described. There are no long-term quantitative studies comparable to those made on trees. We can only speculate on mechanisms responsible for their diversity. Assuming that the intermediate disturbance hypothesis applies to them, we could postulate that high diversity would be largely maintained by a high degree of "openness" and "nonequilibrium" that is even greater than in the large tree communities. The strength of disturbances is scale-dependent. A lightning strike may clear an area of some square meters and enable early tree colonizers to invade. Much less is needed to create a "gap" for some forest invertebrates, for example land planarians. For such species, even a branch or a few leaves fallen from a tree may create "clearings" large enough for species requiring a small disturbed area. A bird dropping may represent an empty gap for some species of insects and microorganisms. Such small disruptive events are much more common than larger disturbances, they are daily or even hourly events. So, it seems likely that nonequilibrium conditions are the rule for communities of small organisms: there will always be an oversupply of empty gaps ready for colonization, and there will always be numerous local communities at different stages of succession, resulting in great diversity. Future studies should attempt to analyse such events quantitatively. Advantages for such studies include the relatively small areas necessary for assessment, and the shorter time scales. Difficulties include the taxonomic diversity and scanty knowledge of the flora and fauna.

In conclusion, agreement about the relative importance of various factors contributing to the maintenance of the great species diversity of trees in tropical rainforests has not been reached. However, it seems likely that a range of mechanisms is important, including repeated environmental disruptions, spatial and temporal heterogeneity, operation of compensatory mechanisms that prevent displacement of rarer species, and gradual climatic change, among others. It may well be that communities of small species, whether plant or animal, which are vastly more speciose than trees, are controlled by a different mix of factors. Quantitative studies are necessary to provide evidence. Finally, with the exception of the brief consideration of coevolution (which is not likely to be of any importance, at least for trees), the discussion was entirely concerned with the *maintenance* of diversity, but what about its *evolution*? Is it possible that a factor not even mentioned in the various studies, i.e., direct temperature effects on evolutionary speed, may be the most important? This aspect is discussed in Chapter 9 (pp. 159–165).

Ectoparasites of marine fish: non-interactive unsaturated communities

As mentioned in the section "Species nonsaturation and nonequilibria", parasites of fishes are very speciose and represent a major component of the Earth's fauna. They are ideal objects for ecological studies because they live in well defined habitats, the number of possible replicas is practically unlimited, and the distribution of parasites in their microhabitats can be easily mapped and quantitatively evaluated. Rohde and collaborators have studied extensively the ecological and zoogeographical aspects of a wide range of ectoparasites of marine fishes. Important taxa infecting the gills, mouth cavity, fins and body surface of marine fish are, in order of importance, copepods, monogeneans, isopods, and trematodes. Comparison of the numbers of ectoparasite species infecting different fish species leaves no doubt that many vacant niches exist, even in fish that are heavily infected with a great number of parasite species (e.g., Rohde 1979a, 1998a). Thus, the gills and head of *Lethrinus miniatus*, a fish of medium size from the Great Barrier Reef, Australia, is heavily infected (up to about 3500 parasites of all species per host) with about 25 species of metazoan ectoparasites, mainly monogeneans and copepods. Fish were examined at three localities. Maximum prevalence of infection with a particular parasite species reached 100 %, and the number of parasite species per host was 5–11. But the bony gill arches, infected with various parasites in other fish species, were never infected (Rohde *et al.* 1994).

5666 fish of 112 species had an average of 4.3 species of metazoan ectoparasites on the heads and gills per fish species (Rohde 1998a). The maximum number was 27 on *Acanthopagrus australis*, a small fish from warm-temperate waters in southeastern Australia. The vast majority of fish species had fewer than 7, and 16 had none (Figure 2.6A). Assuming that 27 species is the maximum a host species can support (and there is no reason for this assumption: pathological effects are minimal) and that other fish species could support the same number (and there is no reason why they should not), maximally only 15.9 % of all niches are filled. Similarly, considering abundances (the total number of all parasites of all species per number of fish of a particular species examined), maximum abundance was about 3500, but most fish had an abundance of fewer than 5 (Figure 2.6B). The mean abundance of 54.68 represented 1.82 % of the maximum, again indicating that many more parasites could be accommodated.

Different parasite species may be dominant, either by intensity of infection or biomass (volume), on different host individuals, probably depending on which infective stages are most common in particular habitats, or on behavioral differences between fish which determine contact with particular parasite species (e.g., Rohde *et al.* 1994) (Figure 8.1). There is no evidence that the dominance pattern is somehow determined by a hierarchy of interactions, although the occurrence of some interactions cannot be excluded.

Microhabitats of parasites may be relatively large (e.g., all gills) or very narrow (e.g., a small portion of the gills). Microhabitat preferences are genetically determined, i.e., even in the absence of competitors and

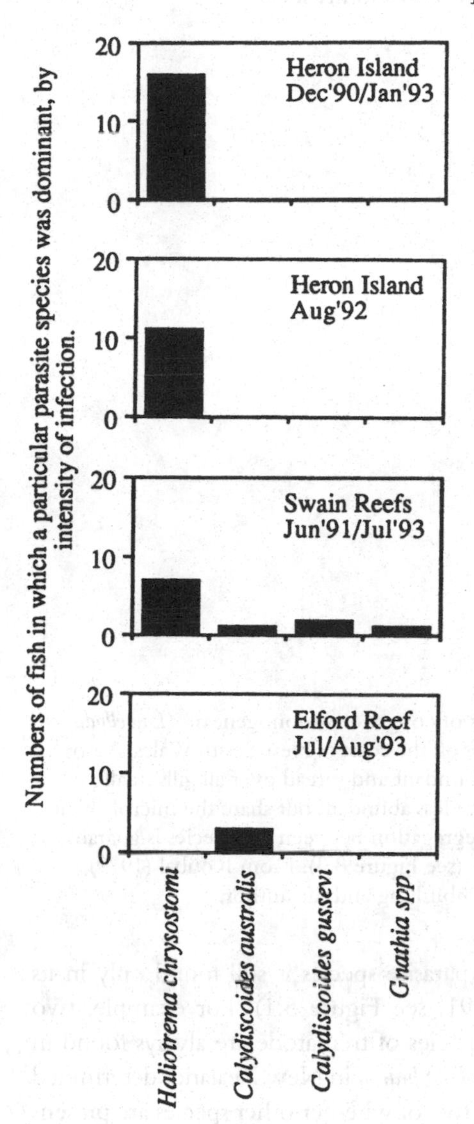

Figure 8.1. Dominance patterns of metazoan ectoparasites on a tropical fish, *Lethrinus miniatus.* Ordinate: number of fish in which a particular parasite species was dominant by intensity of infection. Note: different species are dominant on different reefs. An even greater variability is found if dominance is calculated by volume of parasites. From Rohde, Hayward, Heap, and Gosper (1994), reprinted with permission from Elsevier.

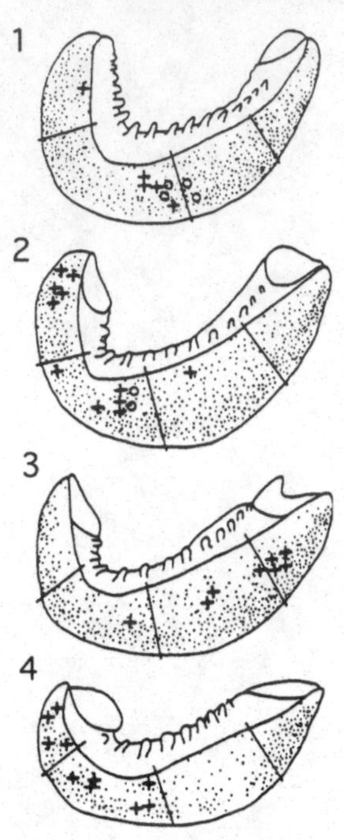

Figure 8.2. Distribution of three species of congeneric monogeneans (*Lamellodiscus* spp.) on the gills of *Acanthopagrus australis* off the coast of New South Wales, Australia. Note: one species (small dots) is very abundant and spread over all gills in all microhabitats. The other two species are less abundant but share the microhabitat with the other species. Reproductive segregation between the species is guaranteed by differently shaped copulatory organs (see Figure 8.3). From Roubal (1979), reprinted with permission of CSIRO Publishing and the author.

otherwise empty gills, a particular parasite species is still found only in its characteristic site (e.g., Rohde 1991, see Figure 5.1). For example, two species of monogeneans and one species of trematode are always found in their particular microhabitat of *Seriolella brama* in New Zealand, determined by their attachment organs, irrespective of whether other species are present or not (pp. 29–31, Figures 2.1–2.3). Occasionally, microhabitat width may expand somewhat in heavy infections, or it may change slightly due to effects of other – possibly competing – parasite species, but often there is no change whatsoever (e.g., Hayward *et al.* 1998). In some hosts, certain parasite species use niches with little or no overlap, in others many species co-occur in the same microhabitat. Comparison of niche overlap of congeneric and non-congeneric monogeneans has shown that congeners tend to overlap less. Congeners that differ in the shape and size of copulatory organs often

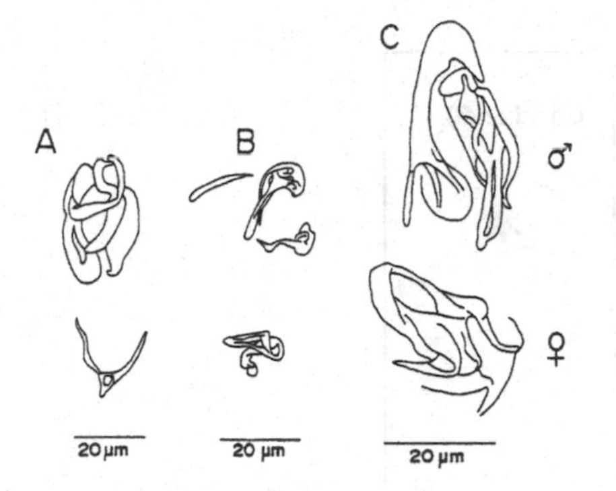

Figure 8.3. Male and female copulatory sclerites of three species of *Lamellodiscus* inhabiting the same parts of the gills of *Acanthopagrus australis* (see Figure 8.2). From Roubal (1979), reprinted with permission of CSIRO Publishing and the author.

share the same niche, whereas congeners with identical copulatory organs are always completely or almost completely segregated into different niches, strongly suggesting that reinforcement of reproductive barriers and not interspecific competition is responsible for niche segregation (Rohde and Hobbs 1986) (Figures 5.2, 8.2, 8.3). Monogenea use blood and mucus/epithelial cells as food, which are fast replaced and not in limited supply as long as the fish is alive; the only potentially limiting resource is space for attachment. Nevertheless, different species of Monogenea, some of them congeneric, on the gills of the same host species differ in the size and shape of their feeding organs such as pharynx and oral suckers (Figure 3.1). This suggests that differences in feeding organs may be fortuitous (e.g., Rohde 1991) and not due to competition for different food resources, as frequently suggested for other animal taxa by various authors. Nevertheless, the possibility cannot be excluded that differences in feeding organs of monogeneans have an adaptive value: they may permit extraction of food (blood) from different parts of the gills.

Importantly, metazoan ectoparasites (and endoparasites) of fish do not conform to the packing rules of Ritchie and Olff (1999) (Figure 2.8), strong evidence that ectoparasite communities are not densely packed and that competition for limiting resources has not been important in evolution. Further evidence against an important role for interspecific competition in ectoparasites of fish is that positive associations are much more common than negative ones (e.g., Rohde *et al.* 1994, 1995), that nestedness is uncommon (Worthen and Rohde 1996; Rohde *et al.* 1998), and – where it occurs – may be the result of epidemiological processes

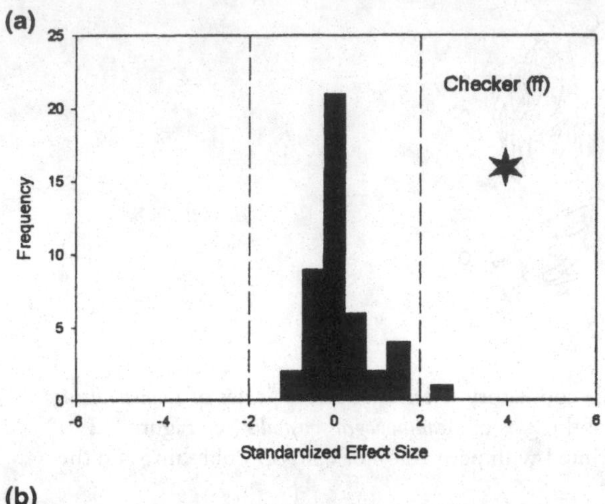

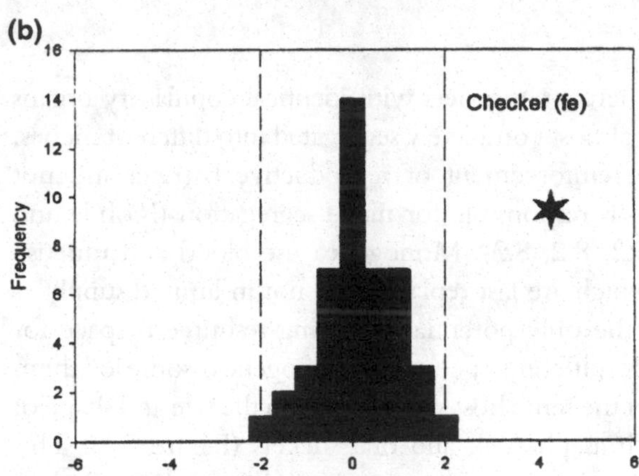

Figure 8.4. A null-model analysis was conducted using ectoparasite communities of 45 species of marine fish. Little evidence for nonrandom occurrence patterns was found. One example is illustrated here. Histograms of SES (standardized effect sizes) of the number of species pairs forming checkerboard distributions are shown. Each observation is an SES for a different host–parasite presence–absence matrix. (for details see Gotelli and Rohde 2002). The broken vertical lines indicate +2 and −2 standard deviations, the approximate boundaries for statistically significant patterns. The null-hypothesis is that the observed distribution does not differ significantly from a mean of 0. 0. The asterisk indicates the tail of the distribution for which species co-occurrence would be less than expected by chance, indicating competitive (?) structuring. (a) Fixed-fixed null-model algorithm; (b) fixed-equiprobable null-model algorithm. From Gotelli and Rohde (2002). Reprinted by permission of Blackwell Science Ltd.

(Morand *et al.* 2002), that there is reduced interspecific relative to intra-specific aggregation (Morand *et al.* 1999), and that null-model analysis found little evidence for nonrandom co-occurrence patterns (Gotelli and Rohde 2002) (Figure 8.4). Finally, very few species of metazoan hyperparasites of ectoparasites of marine fishes have been described, and it is highly unlikely that many exist: further evidence for the existence of many vacant niches and the lack of competition for these niches (e.g., Rohde 1989).

The findings of these studies have been corroborated by several other authors, using marine and also freshwater fish. For example, the recent study of Geets *et al.* (1997) showed that the niche breadth of gill parasites of the fish *Siganus sutor* did not change with increasing abundance of other species present. Simkova *et al.* (2000, 2001a, b, c) have made detailed studies of nine species of the monogenean *Dactylogyrus* on the freshwater fish *Rutilus rutilus*. They found that assemblages of the monogeneans at the local level (among host populations) and at the level of seasons (among host populations within localities) were nested, but that nestedness was uncommon at the level of hosts (infracommunities of parasites). The authors suggest that nestedness may arise from a variety of causes and is not necessarily due to competition. Monogeneans have hard copulatory and attachment sclerites. Simkova *et al.* (2002) measured them in detail, and the attachment sites on the gills for each individual were determined. Morphometric distances of the attachment organs and the copulatory sclerites, and Levin's niche size and Renkonen's niche overlap indices were calculated. The results supported the view that reinforcement of reproductive barriers is responsible for niche segregation, because monogenean species sharing the same niche differ more strongly in the structure of copulatory sclerites than species that are spatially segregated. Furthermore, species with overlapping niches have similar attachment organs, which suggests that morphologically similar species have similar ecological requirements and that interspecific competition is of minor importance, with little impact on the evolution of differences between attachment organs.

In conclusion, evidence is convincing that ectoparasites of fish live in largely unsaturated habitats with many vacant niches. Communities are not significantly structured by interspecific competition.

Insects on bracken, and wasps: type I communities with little evidence for interspecific competition

Among the most thoroughly studied insect communities are insects of the bracken fern (*Pteridium aquilinem*). Lawton and collaborators have studied them for more than 30 years, beginning in Britain and later extending the studies to other countries. Lawton (1999) briefly summarizes his results, and Lawton's (2000) review gives greater detail. In Britain, the fern is extremely common, growing among other plants or in large monocultures, some of which extend over hundreds of hectares. The fern has an underground rhizome and fairly evenly spaced fronds of uniform size, which appear first in April/May and reach a height of more than 1 m. They die with the first frost in autumn. The fronds are a source of food to a variety of insects. At an open site, which was sampled at two- to three-week intervals throughout the growing season in most years between 1972 and 1990, 24 species of insects were recorded, and 27 species were found to regularly feed on bracken foliage in Britain, although some of them were not recorded every year, possibly due to a failure in sampling. A patch at a woodland site was sampled for 8 years between 1980 and 1987. Other patches in Britain were sampled sporadically. The open site had 15–19 (mean 17.6) species of feeding stages per bracken frond, the woodland patch had 3–16 (mean 14.6). There were often unoccupied fronds. Several major questions were asked. The first question relates to the role of density–dependent processes in species dynamics. Statistical tests and consideration of the rank order of species abundances in an assemblage were used to answer the question. Statistical tests revealed direct or delayed density dependence in about three quarters of the species in the open site and in about one third in the woodland site, although the rates may be somewhat lower because of statistical error. Using rank orders of species abundances, i.e., whether common species remained common and rare species rare over considerable periods, tests revealed that rank orders are in fact conserved over some time, although there is a steady decline in W (Kendall's coefficient of concordance). All this means that the community has a regular structure which is slowly changing over time.

A second question relates to the importance of top-down and bottom-up effects in structuring the communities. An attempt was made to answer this question by combining field observations with mathematical models and field experiments. However, a problem arose because experiments on insect communities of bracken were very difficult: larvae of many insect species live in galls, mines or silk-woven refugia that could not be manipulated. It was difficult to make adults lay eggs in field cages, and many species occurred in low population densities leading to logistical difficulties. Nevertheless, some results on the relative contributions of top down effects (by predators, parasitoids and disease) and of bottom up effects (due to resources, especially food) were obtained. They revealed effects of the latter on population abundances at least of some species, i.e., food is not in unlimited supply for some species, and not all fronds are equally suitable as resources, although the reasons for this are not understood and it is not known whether these differences lead to density-dependent limitations in resources.

There are distinct differences in the insect communities between the open and woodland patches, but it proved impossible to determine the reasons. Concerning top down effects, observations showed strong effects of this kind in a sawfly species at the open patch: the species colonized individuals of herbivore-free bracken in an experimental setting, but unaided. The bracken was separated into two groups. In one group, the colonization led to a population explosion in the following years. Fronds were strongly defoliated, as never seen in the field. The population collapsed and disappeared completely, as the result of infection with an unidentified microorganism. In the other patch, the sawfly population persisted and there was no disease before observations ceased a few years later. Lawton concludes that the population explosion was due to release from an unidentified disease, indicative of strong top down effects. Similar top down effects were demonstrated for a delphacid bug. Densities of the bug were raised in experimental areas and controls were kept in unmanipulated areas. There was a rapid (over a few days) convergence of densities both of controls and experimental fronds. Several replicas gave the same results, which were not the result of density-dependent dispersal nor of density-dependent interference. The decline was not due to greater numbers of a predator, a spider. Reasons remain unclear, but parasitoids may be responsible. Lawton concludes that, although results are not "as clean as I would wish," it seems that both bottom up and top down processes are involved in determining abundance, but for most of these processes it is unclear whether they are density-dependent or density-independent. Ants, which feed on bracken nectaries, have no effect on species richness or abundances of insects. Ants feed on caterpillars of other plants, but the bracken caterpillars appear to be protected against ants by special adaptations, such as living in galls, mines, or silken webs, possessing distasteful haemolymph, or using avoidance behavior. To test for interspecific competition, statistical correlations were used, because experiments could not be conducted. Does greater presence of one species depress numbers of others? No such correlations were found.

For the community as a whole, Lawton concluded that, in spite of some ambiguous results mentioned above, most insect species on bracken are regulated by density-dependent processes; bottom up and top down processes affect abundances. There is much predictability in the rank order of abundances, but predictability decreases over time. Nevertheless, insect

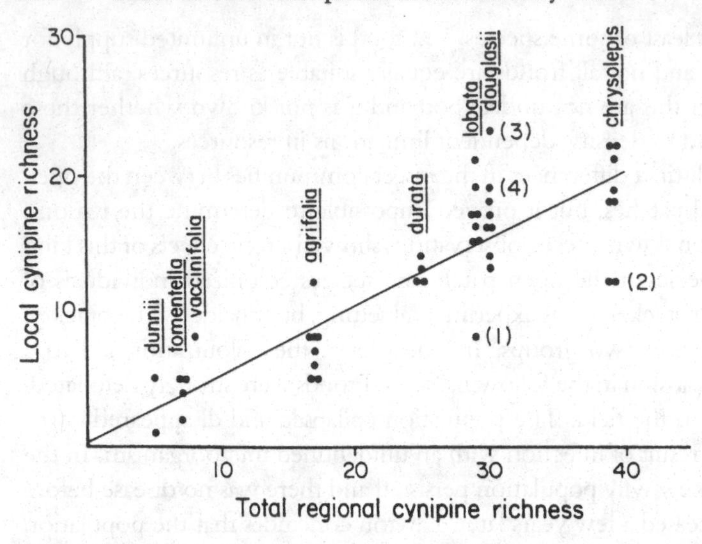

Figure 8.5. Local versus regional species richness of stem-, bud-, and leaf-associated cynipine wasps on Californian oaks. Each point between regional and local richness represents a sampled local host population. Note that even at very high regional richness the relationship does not become asymptotic, suggesting that local communities are not saturated. From Cornell (1985a). Reprinted by permission of the author and the Ecological Society of America.

species tend to respond individualistically to environmental change. Types of species, but not their abundances, are constrained by the need to avoid ant predation. Competition for food is not important. Comparison with other systems shows that each is different; there are useful generalisations, but not at the level of intense local studies.

Studies of bracken insects were extended to other geographic areas, i.e., New Mexico, South Africa, Papua New Guinea, and several British sites. These studies also showed that bracken insects are probably type I communities (see below), although data points are few. The size of the regional pool is determined by how common and widespread bracken is in each geographic area, indicated by a strong linear relationship between area and species numbers. The presence of type I communities corresponds to the finding that interspecific competition was shown to be unimportant (see above). This is further supported by the finding that there are many vacant niches (e.g., Srivastava *et al.* 1997). There are large differences in the numbers of occupied niches between continents, and a particular niche is sometimes occupied by one or many species.

Cornell (1985a, b) censused 7 of the 15 *Quercus* oaks in California for cynipine gall wasps and found a linear relationship between local and regional species richness, given as evidence for nonsaturation of habitats (Figure 8.5). Similar results were obtained by Hawkins and Compton (1992) for African fig wasp communities, both in temperate and tropical habitats. Wasps examined comprised pollinating and non-pollinating gallers, as well as their parasitoids. No evidence was found for saturation in either gallers or parasitoids. For gallers, local species richness was more or less the same at all latitudes (6°N to 34°S), whereas local parasitoid community richness dropped slightly towards the tropics, suggestive of less saturation in the tropics than at high latitudes. In all of these cases, interspecific competition may still occur, but is too weak to result in saturation, supporting the view that herbivore assemblages in general may experience only weak interactions (Lawton and Strong 1981). The finding on parasitoids is of particular importance, since parasitoids comprise about 20–25% of all insect species (Godfray 1994).

In conclusion, interspecific competition for food is unlikely to be important for these apparently unsaturated communities. Although there is much predictability in the rank order of abundances in insect communities of bracken, predictability decreases over time.

Larval trematodes in snails: evidence for interspecific competition (and predation) in infracommunities, and for nonequilibrium conditions

Snails are intermediate hosts for a large variety of trematode species. Because of their small size and the possibility of dissecting large numbers with relative ease, many studies have dealt with the community structure of trematode communities of a range of snail species. Here the groundbreaking studies of Kuris (1990); Kuris and Lafferty (1994); Lafferty *et al.* (1994); Sousa (1992, 1993), among others, must be mentioned. Curtis and Hubbard (1993) examined infections of the 379 littoral snails of the species *Ilyanassa obsoleta* on Cape Henlopen, Delaware, with 5 species of trematode larvae, and found 22 trematode combinations in individual snails; negative effects due to interspecific combinations could not be statistically demonstrated. Kuris and Lafferty (1994) assessed data from 62 studies, including the one by Curtis and Hubbard, in which 62 942 snails out of a total of 296 180 host snails examined were infected. An average of 24% of the snails had multiple infections. Statistical tests showed that of the 14 333 expected double infections, only 4346 were observed. The authors estimated that 13% (10% average) of the trematode infections were lost as the result of interspecific competition (although some of the interactions

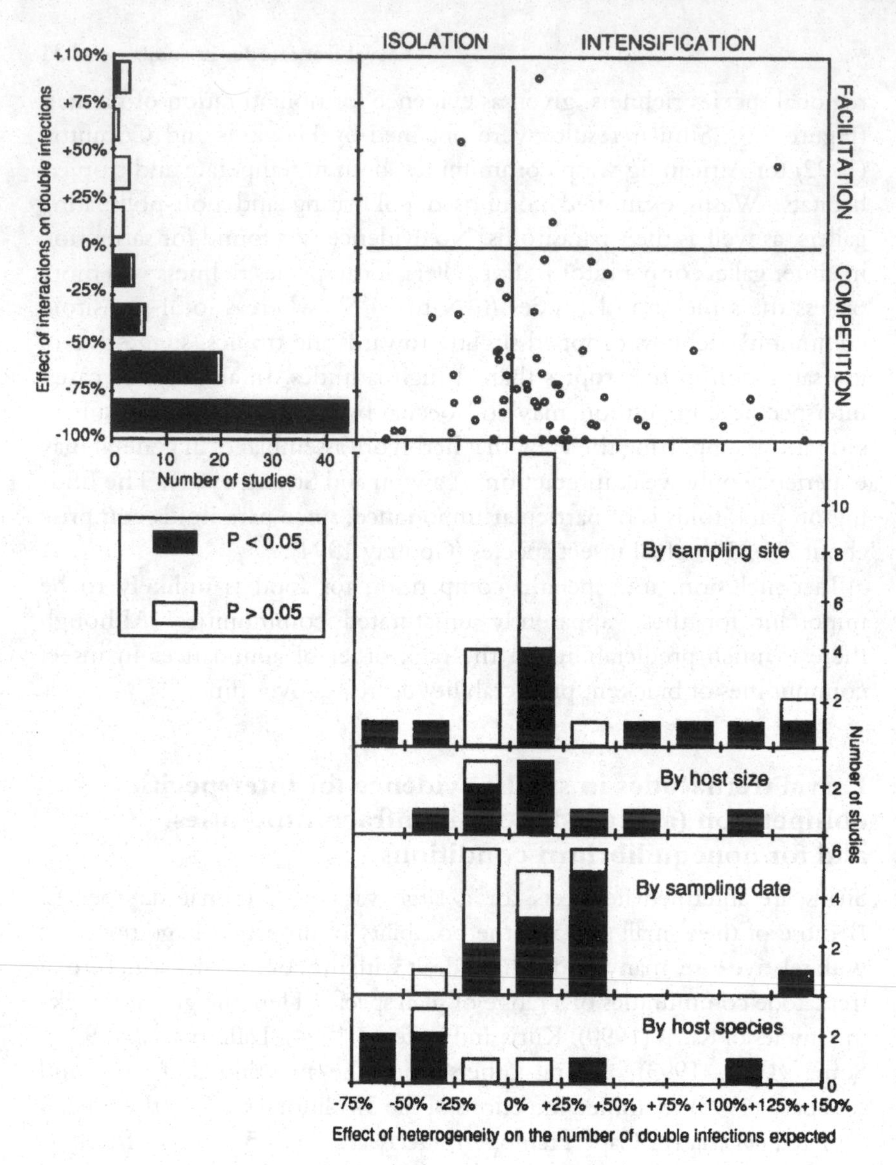

Figure 8.6. Larval trematodes in snails. Effects of interspecific competition and heterogeneity on double infections. Each point in the scatter plot represents a separate study. The effect of interactions on the number of double infections was calculated as (observed – sum of expected)/(sum of expected), whereas the effect of heterogeneity on the number of double infections was calculated as (sum of expected – expected of pooled)/(expected of pooled). A value of zero indicates no effect; a value of –100% indicates complete loss of double infections. Filled bars represent studies in which a significant effect was detected; open bars represent studies where the effect was not significantly different from zero. In the lower part of the Figure, several studies were subdivided according to categories of sampling site, host size, sampling data and host species. From Kuris, A. M. and Lafferty, K. D. (1994). Reprinted by permission of the authors and the *Annual Review of Ecology and Systematics*.

were in fact predatory rather than competitive: rediae feeding on larvae of other trematode species). Importantly, spatial and temporal heterogeneity and differential prevalences among host size classes, in most but not all cases, did not reduce but increased the likelihood of multiple infections (Figure 8.6). This result is not unexpected because vertebrates that are final hosts to different trematode species are often attracted to the same sites. The few exceptions are a consequence of hosts being attracted to different sites. In spite of these seemingly clear results, the authors caution "Despite the large number of studies encompassed in our analysis, we feel that the definitive study of trematode communities has yet to be done. Future analyses of the effects of recruitment and post-recruitment contributions to community structure should include evaluation of the impact of heterogeneity in snail densities at different sites" by weighting samples according to density. Such an evaluation was performed by Lafferty *et al.* (1994) in their analysis of trematode communities in *Cerithidia californica*, a Californian salt marsh snail. They demonstrated that interspecific competition is indeed the most important structuring force in trematode communities of this snail, and again, spatial heterogeneity was found to increase the likelihood of co-occurrence of trematode species. It is also important to note that interactions occur in a hierarchical order, i.e., certain trematode species are better survivors in multi-species infections than others (e.g., Sousa 1992, 1993; Kuris and Lafferty 1994), and that interactions may be positive, i.e., infection with one trematode species may facilitate infection with another. An extreme case was described by Walker (1979): the schistosome *Austrobilharzia terrigalensis* could only infect snails that were already infected with another trematode species.

Importantly, in spite of the distinct interspecific effects at the level of rich infracommunities, that is, within host individuals, discussed above, effects at the level of host populations appear to be minimal. Sousa (1990) based this conclusion on the findings that, in the host-parasite system studied by him, species richness and diversity of trematodes increased with snail size, i.e., complete dominance by a few species did not develop, and parasite species accumulated with time. Furthermore, neither numbers of uninfected hosts nor variation in host size was correlated with parasite diversity. Any reductions as the result of competition were "more than compensated for by increases in both the number and equitability of other parasite species in older host populations." Overall, no evidence for equilibrium conditions was found at the levels of infra- and component communities.

Rich trematode faunas in snails are common; for example, Kube *et al.* (2002) found 10 species in the mudsnail *Hydrobia ventrosa* in coastal lagoons in the southern Baltic Sea, and many other examples are listed in Kuris and Lafferty (1994). Nevertheless, these findings should not create the impression that snails are always filled to such a degree with parasites that multi-species infections and, as a result, interspecific competition are common. For example, the two most common snail species, *Cerithium (Clypeomorus) moniliferum* and *Planaxis sulcatus* on beachrock at Heron Island, Great Barrier Reef, both had a great diversity of trematodes (Cannon 1979; Rohde 1981b), but the less common predatory snail *Peristernia australiensis* whose habitat overlapped with those of these two snails, had no digenean trematodes and only one species of aspidogastrean trematode. Many other rare species of molluscs were never infected at that locality, but results of surveys were not published because snail numbers were too small and negative results are considered to be less interesting.

Farrell (1998) examined *Lymnaea tormentosa*, the intermediate host of *Fasciola hepatica* in northern New South Wales, Australia. Twelve properties, each with a history of *Fasciola* infections in sheep/cattle, were targeted. Population densities of snails were low, but altogether almost 400 snails were examined. None was found to be infected, indicating that even low prevalences of parasites in an intermediate host can keep an infection going, and that great parasite diversities in molluscs are by no means general. Koch (2003) made a survey of all aquatic snails in the same area, the New England tablelands of northern New South Wales. She extensively sampled all nine snail species at a great number of sites and found only two that were infected. One species, *Gabbia vertiginosa*, had three, and another, *Glyptophysa* sp., had one species of larval trematode. Prevalence of infection was low. The tablelands have many ponds and creeks, which are suitable habitats for molluscs, and there is a rich fauna of birds, reptiles, amphibians, and mammals that are potential definitive hosts. However, disturbances due to droughts, and also man-induced disturbances such as pollution due to herbicides and grazing cattle and sheep, frequently reduce population numbers of intermediate and definitive hosts, and are the likely cause of nonequilibrium conditions resulting in low diversity and prevalence of trematode infections.

We conclude that in rich trematode communities interspecific competition is important at the level of infra- but not of component communities. Evidence does not support the view that equilibrium conditions are of any significance. In poor communities (which are common but have been relatively little studied), large and unpredictable environmental conditions (such as rainfall) prevent communities from becoming so rich that interspecific competition can become important.

9 · *Some detailed biogeographical/ macroecological patterns*

Brown (1995) defines macroecology, a term first introduced by Brown and Maurer (1989), as "a nonexperimental, statistical investigation of the relationships between the dynamics and interactions of species populations that have typically been studied on small scales by ecologists and the processes of speciation, extinction, and expansion and contraction of ranges that have been investigated on much larger scales by biogeographers, paleontologists, and macroevolutionists. It is an effort to introduce simultaneously a geographic and historical perspective in order to understand more completely the local abundance, distribution, and diversity of species, and to apply an ecological perspective in order to gain insights into the history and composition of regional and continental biotas." (see also Brown 1999, and Gaston and Blackburn 1999).

Lawton (1999), in a review article entitled "Are there general laws in ecology?" emphasizes the important role of contingencies. There are many widespread, repeatable patterns in nature, but few laws that are generally applicable, because tendencies or rules are contingent on the organisms under study. Such contingencies are observable at all levels, those of populations and ecosystems, but are most complicated at intermediate scales, that is, at the level of communities. In the latter, only "weak, fuzzy generalisations" are possible. Therefore, future research that attempts to dicover general laws, should focus on macroecology, which he defines as the "search for major, statistical patterns in the types, distributions, abundances, and richness of species, from local to global scales, and the development and testing of underlying theoretical explanations for these patterns".

In the following, I discuss some biogeographical/macroecological patterns in some detail in order to demonstrate contingencies for some groups, but also general patterns applicable to several groups.

Island biogeography: evidence for equilibrium conditions?

One of the most widely discussed and controversial issues in biogeography/macroecology is that of species numbers and species dynamics on islands. Islands differ in such features as size, distance from the nearest continent or other colonizing sources, and habitat heterogeneity, and are excellent models to study the effects of these features on diversity and species turnover. Simberloff (1974, 1976), Pielou (1979), Pianka (1983), and Gilbert (1980), among others, give discussions of the equilibrium theory of island biogeography, to which the reader is referred. The main points of the theory are as follows. According to MacArthur and Wilson (1963, 1967), because of balanced immigration and extinction, numbers of species on islands tend towards a dynamic equilibrium; species numbers depend on the area of the island and distance from the source of colonization, with there being a turnover of species once equilibrium has been reached. As pointed out by Gilbert (1980, see also Simberloff 1978 and Connor and Simberloff 1978), the theory as formulated by MacArthur and Wilson does not consider competition between species. When such interactions are incorporated, there would be a noninteractive phase at equilibrium, leading to super-saturation, followed by an interactive phase which either "relaxes" equilibrium or leads to oscillations around the equilibrium and finally to equilibrium (Simberloff). According to Gilbert (1980), demonstration that the theory applies requires demonstration of the following: (1) there must be a close relationship between area and species number; (2) the number of species must remain constant over time; (3) there must be a turnover of species, i.e., some species must be replaced by others over time. The best evidence for this comes from birds. Gilbert is not convinced that the theory describes natural situations well: "many, if not all, insular continental situations are at best badly described by the theory". He believes that a main reason for this is the "extreme oversimplification" of the model, paying no attention to internal habitat diversity and to the differences between species. Therefore, "quantitatively, . . . it would seem that the model has little evidence to support its application to any situation".

However, others have found support for its main assumptions. Wilson and Simberloff (1969) and Simberloff and Wilson (1970) provided experimental evidence for the distinction of noninteractive and interactive equilibrium on islands, by defauning islands and monitoring their recolonization. In the first, noninteractive phase, numbers of species were larger than in the second phase: some species were eliminated by interspecific competition (competitive exclusion), not evident in the first phase.

The experiment described by Simberloff and Wilson was conducted on six very small mangrove islets (11–18 m diameter) in the Florida Keys. The nearest islet was about 2 m, and the most distant one, 533 m from the mainland. The regional species pool of arthropods consisted of about 1000 species, and 20–40 species occurred on each of the islets. Arthropods were extinguished by fumigation, and recolonization was monitored over two years (later extended by Simberloff (1976) to three years). Species reached equilibrium numbers (i.e., numbers before the experiment) in about 200 days, although on two of the islets species numbers remained slightly lower. Importantly, not long after the beginning of recolonization there were slightly higher peaks of species numbers, later levelling off to the equilibrium. The kind of species continued to change, but the total numbers remained more or less constant over the two-year period. The lower numbers on the two islets are unexplained, but may be due to the kinds of species newly acquired, which may have been less well adapted to the habitat, or had different competitive abilities. The islet nearest to the mainland had a distinctly higher equilibrium than the most distant one, the intermediate ones had intermediate equilibrium values. Even after three years, species composition had not converged to the original one (Simberloff 1976).

Moulton and Pimm (1987) studied the effects of the 49 bird species introduced to the Hawaiian islands between 1869 and 1983, and found convincing evidence for competitive exclusion. Pielou (1975) has pointed out that in the interactive phase the equilibrium is dynamic: extinction is continuously balanced by immigration, sometimes by different species. A static quasi-equilibrium will be reached only when an island at the interactive equilibrium level is prevented from further colonization by a suddenly erected barrier.

Also, in contrast to Gilbert's verdict, Rosenzweig (1995) has concluded that the theory of island biogeography has held up well. Island diversity is self-regulating, and competition as well as predation probably helps to increase extinction rates with increasing diversity by negative feedback. However, evidence for the theory comes in the main from some plants and birds.

One of the most important assumptions of the equilibrium theory of island biogeography is that the distance of an island from the nearest colonizing source determines species numbers and rate of species turnover on the island. Obviously, this distance should not be measured in absolute terms but relative to the dispersal abilities and/or vagility of species concerned. Hence, as pointed out above, most evidence for the theory comes from birds and some plants which have great vagility or disperse well. It is unlikely that it will apply to most invertebrates which disperse only sporadically and over short distances. For these taxa, disturbances on islands some distance from a colonizing source that lead to species reductions, will lead to long-lasting nonequilibria. The theory has been extended to make predictions about metapopulation dynamics, host–parasite systems, etc. For example, Nowak and May (1994) and Levin and Pimentel (1981) have used features of the theory and applied them to host–parasite systems. Price (1980), in particular, has provided evidence that most parasite communities are commonly not in equilibrium. It is therefore unlikely that they can be described by the equilibrium theory of island biogeography.

Hubbel's (2001) neutral theory of biodiversity is a generalization of MacArthur and Wilson's theory of island biogeography. Volkov *et al.* (2003) have shown that the relative species abundance of tree species on an island in Panama is better descibed by the neutral theory than a log normal distribution (but see McGill 2003). As shown in Chapter 4 (The packing rules based on fractal geometry and competition) and p. 76, trees (and herbivorous mammals), in their ecological characteristics, are quite different from many other groups, and care should therefore be taken not to generalize these findings.

In conclusion, there is evidence for regulation leading to equilibrium in some island communities (such as birds and some arthropods), but application of the island theory of biogeography to other systems (such as parasites) is controversial.

Inter- and intraoceanic patterns: historical events and centers of diversity are important

Some authors have claimed that similar numbers of taxa have evolved in different large regions, such as continents and oceans, indicative of equilibrium conditions. In the following, I show that species numbers, at least for certain groups, are quite different in large habitats of comparable size and complexity, and that historical events, and in particular evolutionary time available for acquisition of species, are important factors determining extant diversity in particular habitats.

Monogenea and eel parasites in the Atlantic and Indo-Pacific Oceans

Monogeneans are ectoparasitic flatworms inhabiting the gills, fins, and body surface of fish (and occasionally other sites and other hosts). They range in size from less than a millimeter to several centimeters long and are therefore easily detected under the microscope, permitting rapid examination of large numbers of fish. They are thus good models for ecological studies. Rohde (1980c, 1986) demonstrated that relative species diversity (number of parasite species per host species) of monogeneans infecting the gills of marine fish increases from high to low latitudes both in the Atlantic and Pacific Oceans, but diversity is significantly greater in the latter. He suggested that the older age of the Indo-Pacific is responsible for its greater diversity, because numbers of species of Gyrodactylidae, which are predominantly cold-water species and are therefore unlikely to have immigrated from warmer southern waters, are much more numerous in the northern Pacific than Atlantic, and this in spite of the fact that the area of the northern Pacific is not greater than that of the northern Atlantic. Structural complexity, as indicated by the length of the coastline and the number of islands, is also not greater in the northern Pacific, and the lengths of major rivers draining into the northern Pacific and their discharge rates and annual discharge volumes are actually smaller than those of the northern Atlantic, indicating that more gyrodactylids could have immigrated from freshwater into the northern Atlantic than the Pacific, which – however – did not happen.

Another example is parasites of eels. Eels (*Anguilla* spp.) spawn in the ocean, but migrate into rivers and lakes, where they spend much of their lives before returning to the sea. Their parasites may be acquired in freshwater or the sea, although most parasites are acquired in freshwater. Parasites of eels have been studied in detail by Hine (1978, 1980a, 1980b), Hine and Francis (1980), Kennedy (1985, 1990, 1992, 1995), Gosper (1992), Kennedy and Guégan (1994, 1996), Kennedy *et al.* (1997), and Marcogliese and Cone (1993, 1996, 1998). The long association of eels with the oceans (Aoyama and Tsukamoto 1997; Tsukamoto and Aoyama 1998; Aoyama *et al.* 2001; Tsukamoto *et al.* 2002) permits conclusions concerning the effects of history, geographical region, etc. on parasite diversity. Protistans and myxoxoans were not examined by all authors and are therefore not included in this discussion.

Data in Kennedy and Guégan (1996) show that infracommunity richness of intestinal parasites in 1175 *Anguilla anguilla* at one locality in

Britain observed over 17 years never exceeded four species, and that of 1276 eels at 64 other localities in England and Ireland observed over shorter periods never exceeded three. The maximum component community richness was nine species in the long-term study, and eight in the others. Between 45% and 62% of the fish were free from parasites altogether. Infracommunities were composed of different species at different localities. The total number of intestinal species was 19.

The infracommunity richness of intestinal parasites of *Anguilla anguilla* in four Italian lagoons was similar to that in British eels, although the composition of the parasite fauna was quite different. Most eels were not infected or had a single parasite species; the maximum infracommunity richness was three, and the component community richness was eight, four, four, and three respectively (Kennedy *et al.* 1997). Except for one lagoon, where some freshwater parasites were found, all parasites were marine or euryhaline, and digeneans – very rare in Britain – dominated the communities, although few digeneans appeared to be eel specialists. The most prevalent and abundant species were usually specialists.

In all British and Italian localities, the relationship between maximum and mean infracommunity richness and component community richness of parasites was "curvilinear," best described by a power or polynomial function. Kennedy and Guégan (1996) and Kennedy *et al.* (1997) interpreted this as meaning that infracommunity richness became increasingly independent of component community richness, concluding that these findings cannot be explained by "supply side ecology" (i.e., pool exhaustion or transmission rates), but only by processes acting within infracommunities that limit species numbers.

Anguilla rostrata from Atlantic North America (Nova Scotia) also have depauperate parasite faunas, as shown by Marcogliese and Cone (1993, 1996, 1998), who examined a total of 1041 eels from 28 sites in Nova Scotia and recorded a total of 12 metazoan parasite species, including 8 from the digestive tract. Differences between total component community richness, intestinal component community and infracommunity richness between British and Nova Scotian eels were not significant, and eels from both regions were infected with nearly identical suites of eel specialists.

Kennedy (1995; see also Gosper 1992) examined 82 *Anguilla reinhardti* from 10 localities in Queensland (Australia) and recovered 27 species of parasites, 15 of them specific to eels: 17 of the species occurred in the digestive tract, and there were 12 nematode and 8 digenean species. All fish were infected, although 8% were not infected with intestinal helminths. At least 55.5% of the parasite species were eel specialists. It is likely that species richness is even greater, since the Australian studies are based on small sample sizes, and other authors had already recorded one species of monogenean and three species of digeneans from Australian eels, not found by Kennedy. The component community richness of intestinal helminths was 3–9, and the infracommunity richness 1.2–3.6, with a maximum of seven species.

The studies by Hine (1980a, 1980b) and Hine and Francis (1980) of parasites of 839 *Anguilla australis* and 459 *A. dieffenbachi* from 34 localities in New Zealand recorded a total of 21 metazoan species of parasites, of which about 18 are from the digestive tract, including 11 species occurring at high prevalences of infection in both hosts, and 7 at low ($\leq 10\%$) prevalences. Of the latter, one was found only in *A. dieffenbachi*, and four were found only in *A. australis*. No distinction is made between eel specialists and generalists. However, Kennedy (1995) has shown that very few of the rare species and most of the common species in Australia are eel specialists. Therefore, as an approximation, about 11 species can be considered to be specialists; 8 species, including 4 common ones, were trematodes. Data on infra- and component community richness were not given.

According to Marcogliese and Cone (1998), differences in the numbers of specialist species between Britain/Canada on the one hand and Australia on the other are significant: only 7 out of 12 species in Nova Scotia and 8 out of 25 species in Britain were eel specialists, compared with 15 out of 27 species in Australia. This shows that Atlantic eels have much poorer parasite communities than *A. reinhardti* in Australia, particularly with regard to specialist parasites. New Zealand eels share ten helminth genera with Australian eels (Kennedy 1995), but total parasite richness and total richness of intestinal helminths, in particular of species occurring at high prevalences of infection and therefore probably eel specialists, are smaller in New Zealand than in Australian *A. reinhardti*.

How can the differences be explained? The fact that both component communities of parasites and infracommunities are richer in Australia, suggests that "supply side ecology" may be an explanation, and that infracommunities are not limited by competition between species within them, as suggested by Kennedy and Guégan (1996). Rohde (1998b) has used computer simulations to show that a curvilinear relationship between component and infracommunity richness may be a consequence of differential colonization rates and life spans of species in the communities (Figure 4.1). It is not necessary to invoke interspecific competition, although interactions between species and an Allee effect (a reduced survival chance of species occurring in very small densities) may influence the shape of the curve as well. Kennedy (1985, 1992) has shown that some species of eel parasites do indeed interact negatively. Such interactions may modulate the basic shape of the curvilinear curves showing the relationship between infra- and component community richness, but probably in a relatively minor way. The supposition that an Allee effect and not only interactions may be involved, is suggested by the demonstration by Rohde and Hobbs (1988) that simulation of an Allee effect has the same consequence for parasite distributions in host populations as has mortality. Differential colonization rates and life spans are a more parsimonious explanation of curvilinearity than competition and/or Allee effect, because it does not make assumptions about processes at the infracommunity level: the probability of different parasite species all having very high and similar likelihoods of being encountered in communities is infinitesimally small; curvilinearity of graphs representing the relationship between infra- and component community richnesses is therefore expected. Norton *et al.* (2003) also concluded that higher infracommunity richness in eels is possible, and they suggested that infracommunity richness may be a stochastic reflection of component community richness.

Dispersal of eels in geologic time may explain the differences in species richness between Europe and North America on the one hand, and Australia on the other. Aoyama and Tsukamoto (1997), Tsukamoto and Aoyama (1998), Aoyama *et al.* (2001), and Tsukamoto *et al.* (2002) constructed a molecular phylogeny of species of *Anguilla*, and, based on this phylogeny, concluded that ancestral eels originated in the western Pacific (in what is now Indonesia) (Figure 9.1). The species *A. borneensis* in Borneo is the most likely basal eel in the tree. *Anguilla* may have originated ca. 50–60 Ma (during the Cretaceous-Eocene) and, from there, one group dispersed through the Tethys Sea before its closure in the Oligocene (ca. 20–30 Ma), the long-lived larvae probably transported by the global circum-equatorial current. It then split into two species, one colonizing eastern Africa (*A. mossambica*), and one colonizing the Atlantic, and further splitting into the European *A. anguilla* and the North American *A. rostrata. A. reinhardti*, *A. dieffenbachi*, and *A. australis*, among others, dispersed southward over much shorter distance. Snails, the intermediate hosts of trematodes, could not disperse with the eels via the Tethys Sea, explaining the smaller number of trematodes and in

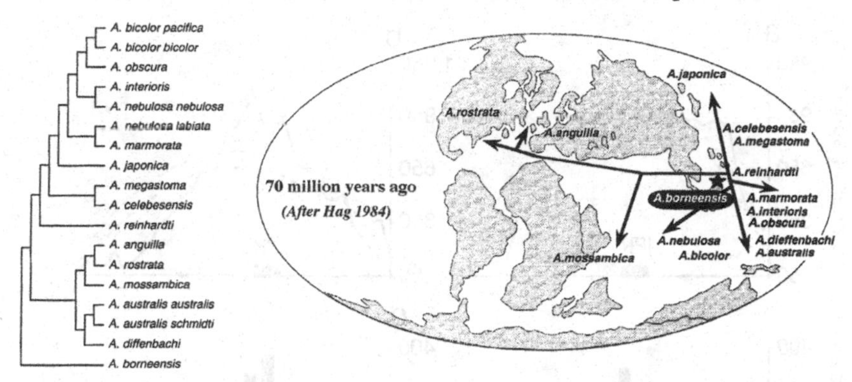

Figure 9.1. Phylogenetic tree and historical migration routes of *Anguilla* spp. modified from Aoyama *et al.* (2001) and Tsukamoto *et al.* (2002). Note: likely origin of *Anguilla* in the western Pacific. One explanation for the reduced diversity of eel parasites in the Atlantic may be a loss of parasites during migration, although a temperature effect (increased species richness at low latitudes) may also be an important factor. With the permission of the authors.

particular of trematodes specific to European eels, compared with *A. reinhardti*. Marcogliese and Cone (1998) also concluded that reduced diversity of parasites of Nova Scotia (Canada) and Europe relative to diversity of Australian eels is due to the shorter time available for the acquisition of (particularly specialist) parasites in the Atlantic. Nevertheless, the tropical/subtropical habitats in Australia versus the cold-temperate ones in Canada and Europe may also play a role, as pointed out by Kennedy (1995).

In summary, then, richness of parasite communities of *Anguilla* spp. cannot be explained by the assumption that a limit to diversity has been reached, determined by the number of species that can be accommodated, and that this limit differs between regions; it can only be explained in two ways, i.e., firstly by historical contingencies, and secondly by temperature differences in the various regions. Concerning the first point, species richness in Atlantic eels may have been reduced as the result of very long distance dispersal over evolutionary time, and, given enough evolutionary time, a greater diversity is likely to develop at least in Atlantic eels, but possibly also in Indo-Pacific ones. Concerning the second point, as pointed out by Kennedy (1995), differences in temperature may be important: a greater species pool in the tropics may have led to an increase in the richness of eel parasites, at least of generalist species (see Chapter 9 (Latitudinal diversity gradients)).

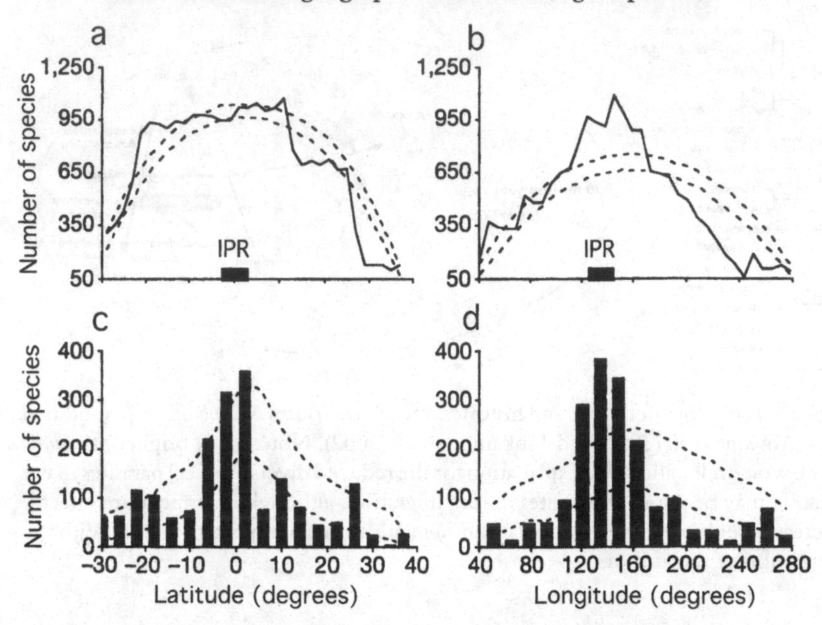

Figure 9.2. Geographical patterns in reef fish diversity in the Indo-Pacific oceans. Richness is far greater in the center than in outlying regions. IPR – Indonesian/ Philippine region. a, b, clines (solid lines) defined as the number of species whose geographical ranges include a point in latitude or longitude, respectively. c, d, distributions of mid-latitudinal or mid-longitudinal ranges (bars). Effects of the mid-domain effect were tested by running a null-model (for details see Mora *et al.* 2003); dotted lines correspond to the maximum and minimum values after running the model. The result shows that the great species richness in the center cannot be explained by a mid-domain effect. Reprinted from Mora, Chittaro, Sale, Kritzer, and Ludsin (2003), with the permission of MacMillan Publishing Ltd., and the authors.

Reef fishes: geographical distribution is determined by a center of diversity and dispersal from it

Mora *et al.* (2003) examined the distribution of 1970 species of Indo-Pacific reef fishes (more than 70% of all the species expected in any reef fish community) from 70 locations, demonstrating a center of high diversity in the Indonesian and Philippine regions. Species numbers decreased steadily along latitudinal and longitudinal axes away from this center (Figure 9.2). Although superficially similar to a distribution resulting from a mid-domain effect, the observed nonrandom distribution with the very high species numbers in the center cannot be explained by such an effect.

Three hypotheses to explain the large-scale pattern in reef fish distribution have been proposed, i.e., the center-of-origin hypothesis (the center is a

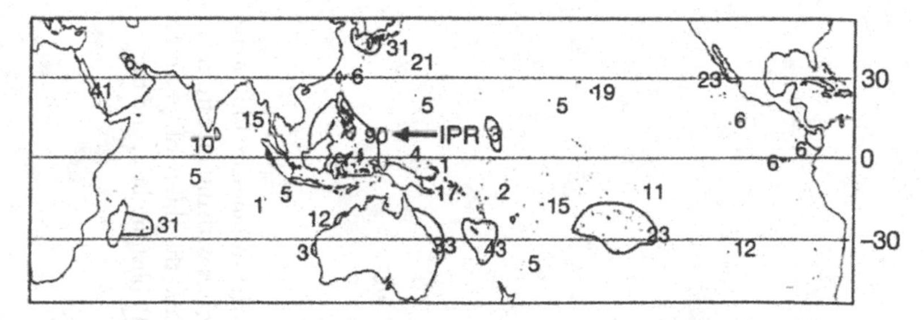

Figure 9.3. Numbers of endemic reef fish (restricted to a single location) in the Indo-Pacific ocean. The center has by far the greatest number of endemic species, supporting the center-of-origin hypothesis explaining the great overall diversity in the center. Reprinted from Mora, Chittaro, Sale, Kritzer, and Ludsin (2003), with the permission of MacMillan Publishing Ltd., and the authors.

center of speciation from which species disperse); the center-of-overlap hypothesis (high diversity in the center is the result of overlapping faunas from more marginal regions); and the center-of-accumulation hypothesis (speciation has occurred in several peripheral regions and ranges have expanded due to prevailing currents into the center, some marginal populations may have been lost secondarily). Data permit a conclusion on which of the hypotheses is correct. The last two hypotheses can be ruled out because peaks of plots of species' longitudinal and latitudinal mid-ranges coincide with the center, and because the center has the greatest number of endemics (Figure 9.3). There are some minor centers of endemism as well, mainly in geographically isolated locations or in locations where the current direction is mainly from tropical to temperate. The great diversity in the center can be explained by the very great number of islands per unit of geographical area, greater than in any other region, facilitating a high level of allopatric speciation, further facilitated by recent geological sea level changes. Species richness decreases, and dispersal ability increases, with distance from the center (Figure 9.4). In other words, community composition is largely determined by the distance to the center and by the ability of species to disperse; locally endemic species contribute little (about 2%) to communities away from the center (Figure 9.5).

In summary, not only latitudinal but also longitudinal gradients in diversity exist, leading to a center of diversity with particularly great richness of endemic species from which outlying regions are colonized. Speciation in outlying regions occurs but contributes little to community richness.

However, oddly, Hughes *et al.* (2002) arrived at quite different conclusions. They found that the centers of diversity do not have the greatest

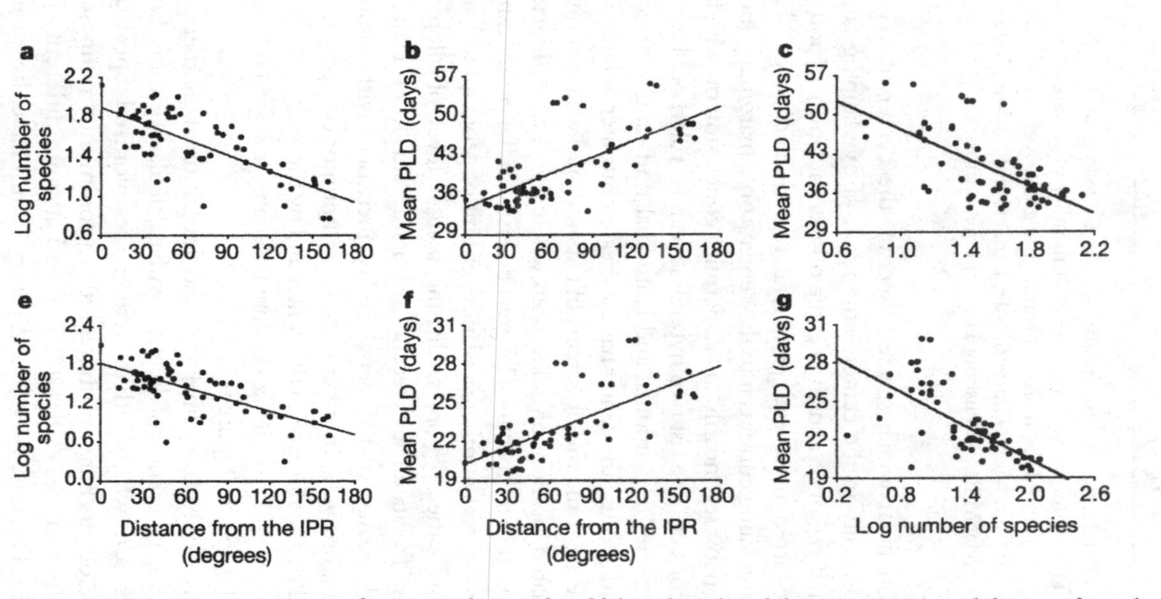

Figure 9.4. Pairwise comparisons of species richness of reef fish, pelagic larval duration (PLD), and distance from the Indonesian/Philippine center of diversity (IPR) for Labridae (a,b,c) and Pomacentridae (e,f,g). The result shows that dispersal ability increases with distance from the center, suggesting that community composition of reef fishes is largely determined by distance from the center and the ability of species to disperse. Reprinted from Mora, Chittaro, Sale, Kritzer, and Ludsin (2003), with the permission of MacMillan Publishing Ltd., and the authors.

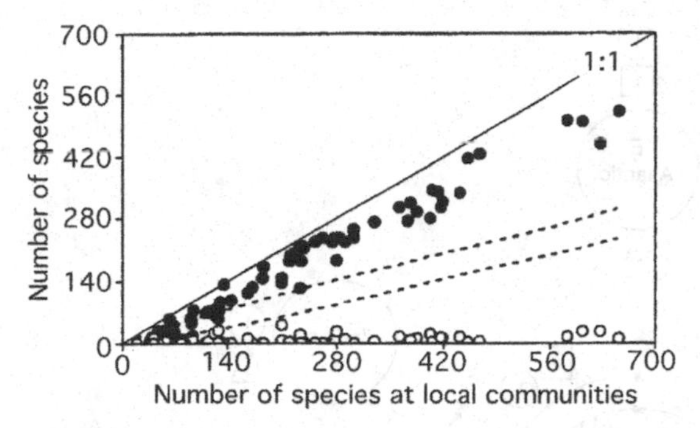

Figure 9.5. Contribution of endemic species of reef fish (open circles) and species from the Center of Diversity (IPR) (filled circles) in the Indian and Pacific Oceans. The number of IPR (Indonesian-Philippines Region) species in communities that can be expected by chance was assessed by determining the number of IPR species in communities (of the same size as observed) randomly generated from the total species pool. Species were selected with equal probability and without replacement. Upper and lower limits of the number of species from the IPR in 1000 iterations are indicated by the broken lines. Note that only about 2% of species were endemics in any community, i.e., very few species are generated in communities away from the center, much less than expected by chance. Reprinted from Nature (Mora, C., Chittaro, P. M., Sale, P. F., Kritzer, J. P. and Ludsin, S. A. Patterns and processes in reef fish diversity. Nature 42, 933-936) 2003, with the permission of MacMillan Publishing Ltd., and the authors.

number of endemics, the reason being that reef fishes (as well as corals) have strongly skewed range distributions. Many species are widespread and the largest ranges overlap, generating peaks in species diversity near the equator and in the Southeast Asian center of diversity.

Scombrid fishes and their ectoparasites: geographical distribution is determined by centers of diversity and oceanic barriers

Using extensive taxonomic revisions of all 18 species of *Scomberomorus* and 2 species of *Grammatorcynus* (Scombridae), their 23 copepod and 17 monogenean species, and their geographical distribution, Rohde and Hayward (2000) have shown that the East Pacific barrier has been the most effective barrier for dispersal of these fishes and parasites, and that the American land barrier has been less effective (Figures 9.6 and 9.7). Reasons are that species could spread between the Indo-West Pacific and Atlantic via the Tethys Sea, and between the western Atlantic and eastern Pacific before the Central American landbridge was established, whereas the eastern Pacific has always been an effective barrier to dispersal, even more so in the geologic past than at present. This led to significant

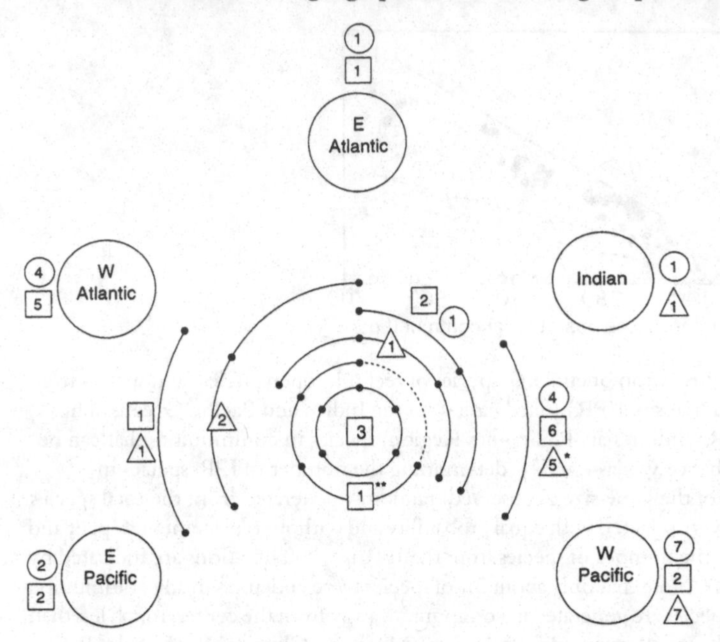

Figure 9.6. Geographical distribution of 18 species of *Scomberomorus*, 2 species of *Grammatorcynus*, 23 species of Copepoda and 17 species of Monogenea. Species numbers restricted to certain seas are given beside the large circles which indicate different seas, numbers of species shared by different seas are given near or on the lines that indicate sharing. Circles: fish; squares: copepods; triangles: Monogenea. Note: the Indian Ocean is defined as not including the coastal zone along the west coasts of Malaysia, Indonesia and Australia. Note: one copepod species only recorded from Port Elizabeth, South Africa, but not from the rest of the Indian Ocean, hence connection between W. Pacific and Indian Ocean indicated by dotted line. The result shows that not a single species (with the exception of four circumtropical ones) is shared by the eastern Pacific and western Pacific. Reprinted from Rohde, and Hayward (2000), with permission from Elsevier.

differences in diversity between oceans. Species of *Scomberomorus* and *Grammatorcynus* in seas of low diversity have relatively more parasite species than those in seas of high diversity. However, when considering only endemic fish and parasite species, the two centers of diversity, i.e., the Indo-West Pacific and the West Atlantic, have parasite/host ratios (= relative species richnesses) of 1.75 and 1.25, respectively, versus a ratio of 1.00 for all other seas. These differences in relative species richness in different seas suggest that more parasite species could be accommodated at least in the poorly infected fish; the differences in relative species richnesses of endemic and all parasite species suggest that historical factors are responsible for the differences. Apparently, fish in low-diversity regions have acquired

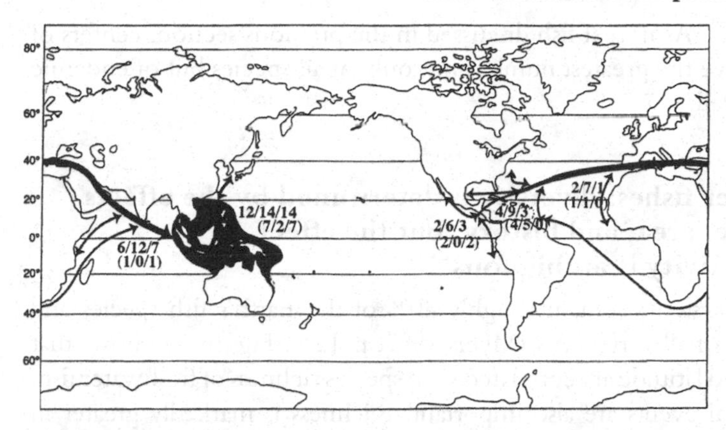

Figure 9.7. Extant geographic relationships of the scombrid fishes *Scomberomorus* spp. and *Grammatorcynus* spp. and their copepod and monogenean parasite species. Numbers indicate species of fish/species of copepods/species of monogeneans (in parantheses: only endemic species). Note the primary center of richness in the tropical western Pacific, and a secondary center in the tropical western Atlantic, as well as the effective barrier to historical dispersal represented by the eastern Pacific. Reprinted from Rohde, and Hayward (2000), with permission from Elsevier.

some parasites from fish in high-diversity regions, which markedly increased the relative species richness of parasites of these fish, whereas acquisitions in the opposite direction were very rare. Considering the relative species richness of individual fish species, the trend is even clearer: the fish species with the greatest number of ectoparasite species has five times more parasite species than the poorest ones, and even when fish of similar size and geographical (latitudinal) ranges are compared, species richness differs by a factor of more than two. This can only be explained by historical events. In other words, there is no evidence that parasite species numbers are, in some way, determined by different carrying capacities of fish species for parasites, or by environmental factors.

A word of caution is necessary, however. Many ectoparasites of fishes have a high degree of host specificity, which probably makes historical events particularly important. Many endoparasites are less specific and may be acquired more easily after host migration into new habitats. This was shown for endoparasites of the rainbow trout, *Oncorhynchus mykiss*, and the brown trout, *Salmo trutta.* Populations introduced into new habitats had no fewer parasite species than those in the original locations, and diversity was even perhaps slightly greater (Poulin and Mouillot 2003).

In summary, historical events, such as the erection and disappearance of barriers, are important in determining richness of marine fish/parasite communities. As in reef fish discussed in the previous section, centers of diversity have the greatest number not only of all species but of endemic species as well.

Freshwater fishes: diversity is determined by the effects of latitude, area, and history, but the effect of productivity is ambiguous

Freshwater systems contain roughly 40% of the Earth's fish species and almost 20% of all vertebrates (Myers 1997, p. 126). Figure 9.8 shows that both area and latitude are correlated with species richness of freshwater fish, but historical events are also important. Richness is markedly greater in North America than in northern Eurasia, although both regions are at roughly the same latitude. The difference can be explained by the presence of refuges during the Ice Ages in North America but not in Eurasia, permitting the survival of more species in the former than the latter (Tonn *et al.* 1990). Madagascar is tropical and has large freshwater bodies (600 000 ha), but its diversity is very low (even when marine species regularly or sporadically invading freshwater are included), probably because of its long (Pre-Tertiary) isolation from larger continents (Kiener and Richard-Vindard 1972).

Concerning fishes in freshwater lakes, the greatest numbers of species are found in some large African tropical lakes. For freshwater lakes, productivity is generally higher at low than at high latitudes, although there is much overlap (net primary productivity of tropical lakes 100–7600 mg C/m^2 per day, 30–2500 g C/m^2 per year; temperate lakes 5–3600 mg C/m^2 per day, 2–950 g C/m^2 per year) (Likens 1975, p. 192). Lakes Victoria, Tanganyika and Malawi (121 500 km^2) have a total of about 1450 freshwater fish species (17% of the Earth's total) (Myers 1997, p. 127). Of these, Lake Malawi (28 231 km^2) contains at least 550 fish species. In contrast, the North American Great Lakes (246 900 km^2) contain 173 fish species (Myers 1997, p. 127), Lake Baikal has 39 (Sheremetyev, personal communication). Lake Baikal and the North American Great Lakes together contain 31% of the Earth's freshwater, but have only 212 fish species. Annual productivity (g C/m^2 per year) compares as follows: North American Great Lakes, *c.* 100 to 310 (Wetzel 1975, Table 14.10), 80–90 to 240–250 (Likens 1975 pp. 194–195); Lake Baikal, 122.5 (Likens 1975, p. 194). Lake Victoria, in contrast, has an annual production of 640 (Wetzel 1975, Table 14.10), which is among the highest for freshwater lakes on Earth. Hence, at least for these

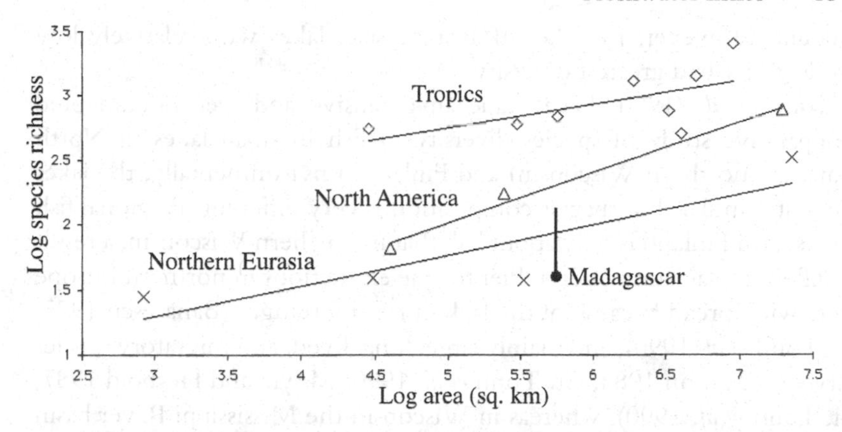

Figure 9.8. Effects of latitude, area, and continent (history) on richness of fish species in some freshwater systems. From left to right, Tropics: Lake Malawi, Malaysia, Thailand, Indonesia excluding Papua New Guinea, India, Congo basin, Amazon basin, South and Southeast Asia. North America: NW Wisconsin, North American Great Lakes, North America. Eurasia: Lake Constanze, Lake Baikal, Finland, former USSR. Madagascar (strictly freshwater species indicated by a filled circle, bar represents number of marine species regularly or sporadically invading freshwater). Note the generally increased diversity in the Tropics, the effect of area, the greater richness in North America than in Europe, and the abnormal species poverty in Madagascar, likely to be due to the long isolation from a large continent. Data for Amazon and Congo basins from Rosenzweig and Sandlin (1997). Data for Madagascar from Stiassny and Raminosoa (1994), for Finland and NW Wisconsin from Tonn *et al.* (1990). Data for Lake Constanze from Geller and Güde (1989). Data for USSR and Lake Baikal from Sheremetyev, personal communication. All other data from various references in Rohde (1997, 1998a).

freshwater fishes, it seems that highest diversity is found in tropical lakes with highest productivity (although data are insufficient to correct for area along a latitudinal gradient). This, of course, does not "prove" that productivity "explains" latitudinal gradients in species diversity. There are other covariables, such as temperature, solar radiation, and seasonality, that may be responsible. Also, as pointed out by Rohde (1998a), high productivity may well be the result rather than the cause of great species diversity.

Dobson *et al.* (2000) have investigated the relationship between primary productivity and species richness in 33 North American lakes. They considered diversity of phytoplankton, rotifers, cladocerans, copepods, macrophytes, and fish. Primary productivity was measured as g C/m^2 per year, ^{14}C estimate. All taxa showed a significant quadratic response to increased annual primary productivity, when lake area was taken into account. However, for lakes of average size, lakes with relatively low productivity had greatest diversity.

Tonn *et al.* (1990) have made an extensive and well documented comparative study of species diversity of fish in small lakes in North America (northern Wisconsin) and Finland. Environmentally, the lakes are quite similar, but species composition is very different. Regional fish diversity in Finland is only about half that in northern Wisconsin, a result of different glaciation effects. Pleistocene extinctions in northern Europe were widespread because of the lack of nearby refuges (Banarescu 1975, cit. Tonn *et al.* 1990), and mainly large, long-lived, and migratory species survived (Mahon 1984, cit. Tonn *et al.* 1990; Moyle and Herbold 1987, cit. Tonn *et al.* 1990), whereas in Wisconsin the Mississippi River basin provided a nearby refuge and center of speciation (Burr and Page 1986), favouring species with short life spans and narrow habitat ranges (Mahon 1984, cit. Tonn *et al.* 1990). In other words, North American fish species are more specialized (references in Tonn *et al.* 1990). Tonn *et al.* demonstrated the following differences for the fish assemblages studied by them. In Wisconsin, assemblages were of the presence–absence type, i.e., small fish prey species were absent from lakes containing piscivores, and there was habitat specialization. In contrast, in Finland, assemblage types could be distinguished only on the basis of relative abundance of species. Plotting local against regional diversity revealed an asymptotic relationship in Wisconsin; in Finnish lakes the asymptotic part of the regression line was absent. This may mean that competition may have limited species numbers in Wisconsin local communities, but the authors stress that isolation, ecological impoverishment, and habitat severity of small forest lakes and size-limited predation are probably more important. "Currently, a general explanation accounting for various local–regional relations cannot be provided."

In conclusion, historical events such as glaciations, long isolation of islands from a large continent, and latitude and area, are important in determining richness of freshwater fish communities, but evidence for the role of productivity is ambiguous.

Latitudinal diversity gradients: equilibrium and nonequilibrium explanations

Anybody who has visited a tropical rainforest and a temperate forest, anybody who has dived on a coral reef and a cold-water shore, and anybody who has visited a tropical fishmarket and one in a temperate climate, will be struck by the enormous differences in species diversity. Numerous studies deal with such latitudinal gradients in species diversity and there can be no doubt that such gradients exist for many taxa and habitats, but there are exceptions (for recent reviews and important papers see Rohde 1992, 1999; Willig 2001; Willig *et al.* 2003; Hillebrand 2004). Also, a gradient in species richness does not always lead to a gradient in community richness (Rohde and Heap 1998). Fossil evidence indicates that the existence of the gradients is time-invariant, although both latitudinal and longitudinal gradients in diversity have strongly increased through the Cenozoic, i.e., during the last 65 million years (Crame 2001). A consensus has not been reached about the causes, although a considerable number of studies have shown that latitudinal gradients in temperature (or other measures of solar energy input, such as annual potential evapotranspiration, solar radiation, actual evapotranspiration) are best correlated with latitudinal diversity gradients of many taxa, and in many habitats (e.g., Currie 1991; Currie and Paquin 1987, further references in Rohde 1992, 1999). For angiosperms, Francis and Currie (2003) recently demonstrated the primary importance of mean annual temperature (or annual potential evapotranspiration) and annual water deficit, and their interaction for global diversity gradients. It is likely that – in view of the generality of the gradients – some primary cause or causes must be involved; this may be solar energy input (temperature) (e.g., Rohde 1978a,b, 1992, 1999). However, several other factors, whose relative importance differs between taxa and habitats, contribute as well. Among the other important factors that have been suggested are area (e.g., Rosenzweig 1995), heterogeneity of the habitats (e.g., Rahbeck and Graves 2001), productivity (e.g., Rosenzweig 1995), and narrower niches in the tropics (e.g., Ben-Eliahu and Safriel 1982). I will give some examples that show that at least some of these other factors are indeed important.

Area

Rosenzweig (1995), following Terborgh (1973), claimed that greater species diversity is a consequence of the larger areas in the tropics. That area is important in determining species richness is, for instance, shown by the recent study of Valdovinos *et al.* (2003). Analyzing the distribution of 629 species of molluscs along the Pacific coast of South America, they demonstrated that richness increases sharply south of 42 °S where shelf area also increases sharply (Figure 9.9). This increase is not the result of

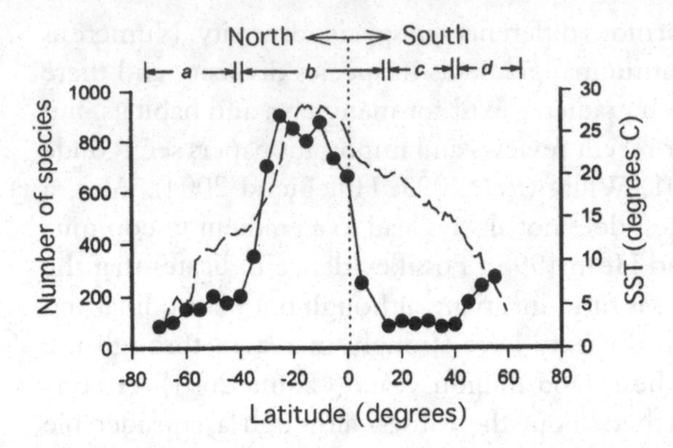

Figure 9.9. Latitudinal gradient in species diversity of molluscs on North- and South American Pacific shelves. Points are selected in bands of 5° latitude. Mean latitudinal SST (sea surface temperature) indicated by segmented line. Note greatest diversity north of the equator, and a rise in diversity in the southernmost zone. From Valdovinos, Navarete, and Marquet (2003). Data for North American molluscs from Roy, Jablonski, and Valentine *et al.* (1998). With the permission of the authors and the Editor of Ecography.

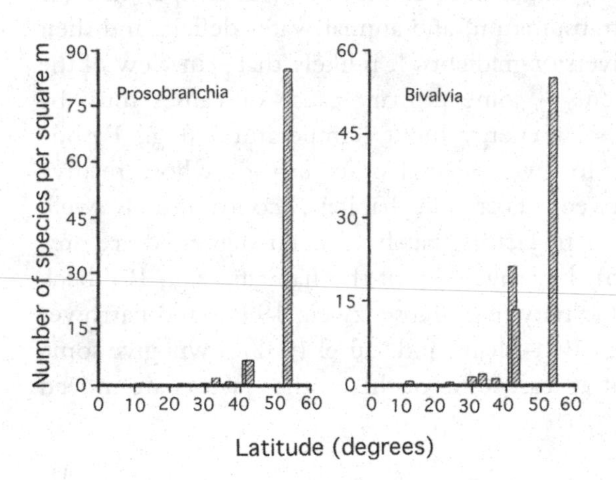

Figure 9.10. Local (alpha) diversity of prosobranch snails and bivalves in shallow soft bottom habitats. Note steep increase towards southern high latitudes. From Valdovinos, Navarete, and Marquet (2003). With the permission of the authors and the Editor of Ecography.

artefacts produced by sampling in the larger area, but is due to an increased alpha diversity, as shown by the results of sampling in soft bottom habitats along the South American coast (Figure 9.10). The authors suggest that geographic isolation due to divergence of ocean currents and the formation of refuges during glaciations facilitated diversification. In contrast, more northern areas have much narrower shelves. Nevertheless, richness in the southernmost areas is much reduced in comparison with the tropical zone, and in North America, both on the Atlantic and Pacific shelves, the latitudinal gradient is more typical with lowest diversity at high latitudes (Roy *et al.* 1998), although the peak of diversity is displaced somewhat to the North.

Figure 9.8 shows the effect of area on species richness in freshwater fish in tropical and cold-temperate waters; the increase is apparent both in cold and warm environments, although warm-water fish are much more diverse than cold-water ones. Other examples of the effects of area on diversity are given by Rosenzweig (1995). So, area is important, but Rohde (1997, 1998a) has given reasons why area cannot give a general explanation of the latitudinal gradients in diversity. These reasons are: (1) tropical land masses are greater than cold-temperate ones only in Africa, in Eurasia nontropical land areas are much larger than tropical areas; (2) many groups of animals and plants have their greatest diversity in Southeast Asia, with a relatively small land area and small freshwater systems (e.g., freshwater fish, for details and references see Rohde 1997, 1998a); (3) the tropical Atlantic has a smaller area than nontropical zones, but nevertheless the greatest diversity in planktonic Foraminifera (reference in Rohde 1997); (4) species richness of many other taxa peaks in the tropics, even if the tropical area is much smaller than high-latitude ones (examples in Rohde 1998a).

Productivity

Rosenzweig and Sandlin (1997) suggest that the reason why some larger areas are so much poorer than small tropical ones is productivity. However, even a cursory glance at some large-scale patterns reveals that productivity cannot be a major factor responsible for the latitudinal gradients. The Antarctic and Subantarctic Ocean is not only a region of large area, but also of very high productivity, producing krill, the staple food of many whales, but diversity is low. On the other hand, most tropical seas do not have particularly high productivity but are very rich in species (for details see Rohde 1998a). Waide *et al.* (1999) have evaluated data from many plant and animal studies and shown that the relationship between diversity and productivity depends on scale. At local and landscape scales (maximum 200 km), absence of, or negative correlations are prevalent; at continental to global scales (more than 4000 km), there are often unimodal or positive relationships, but absence of, or negative correlations occur as well. At regional scales (200–4000 km), an absence of, or negative correlations are about as common as unimodal or positive relationships for plants, and somewhat more common for animals. Also, as pointed out by Rohde (1998a), an element of circularity may be involved when considering productivity: increased productivity may well be the result of greater diversity and not vice versa. Thus, highly diverse coral reefs are highly productive, but they have evolved in low productivity seas (Rohde 1998a). Experiments have shown that increasing biodiversity by a factor of two or three increases productivity by the same factor (Kareiva 1994).

Spatial heterogeneity

A very well supported example for the effect of spatial heterogeneity on diversity was recently provided by Rahbeck and Graves (2001). They analysed the geographic ranges of 2869 species of birds breeding in South America, which represent almost one third of the world's bird fauna. Factors considered were the influence of climate, quadrat area, ecosystem diversity, and topography. The analysis was conducted at 10 spatial scales (about 12 300 to 1 225 000 km^2 quadrat area). Regional variability in species richness correlated best with topography, precipitation, topography × latitude, ecosystem diversity, and cloud cover. Ranking of these factors depended on the scale used for analysis. Direct measurement of ambient energy (mean and maximum temperature) were of ancillary importance. *In toto*, humid montane regions near the equator have the greatest diversity. The authors conclude that a synergism between climate and coarse-scale topographic heterogeneity ultimately controls terrestrial diversity from the poles to the equator. It must not be forgotten, however, that temperature – although not giving good correlations with diversity as such – is the primary factor affecting climate.

Distinct latitudinal gradients in marine plankton show that increasing diversity towards the tropics does not depend on an increase in heterogeneity. Indeed, there is no general trend of increasing heterogeneity towards low latitudes. Exceptions are the presence of a greater variety of habitats in tropical mountain ranges, because there we have tropical habitats at the base, to cold habitats near the peaks (e.g., Körner 2000). Indeed, South America is the habitat where the effect of spatial heterogeneity should be most marked, because the Andes extending along the continent provide an increasingly wide range of habitats towards the equator. The great complexity of coral reefs, often used as an example of the importance of heterogeneity for species richness, is the result of, rather than the reason for, great diversity, as pointed out by Rohde (1998a, see above: productivity).

Narrower niches in the tropics

Several authors have suggested that greater species richness in the tropics leads to denser species packing (e.g., MacArthur and Wilson 1967; MacArthur 1972). One important aspect of a species' niche is its latitudinal range. There have been extensive discussions in the literature on Rapoport's rule, which claims that latitudinal ranges of organisms are generally smaller at low rather than at high latitudes (Stevens 1989). There is controversial evidence for the generality of the rule, which will not be discussed here. The reader is referred to the review in Rohde (1999). However, smaller latitudinal ranges in the tropics (in cases where they exist) do not give an "explanation" of greater numbers of tropical species, they could also be the result of faster evolution at low latitudes, which has led to closer species packing and hence smaller niches including latitudinal ranges. Indeed, Rohde (1998a) suggested the existence of "two opposing trends: on the one hand, tropical species should have large latitudinal ranges because temperature conditions in the tropics are uniform over a much larger latitudinal range than at higher latitudes; on the other hand, newly evolved taxa (in particular subspecies, most of them in the tropics) that have little vagility and no dispersal stages, should have narrow ranges because they did not have the time to spread away from their place of origin even within the tropics. So, Rapoport's rule should apply to recently evolved groups with little vagility, it should not apply to older and more vagile groups", but in either case they do not give an "explanation" of the latitudinal diversity gradient. Moreover, Thorson's rule, according to which marine benthic invertebrates in the tropics generally have widely dispersing pelagic larvae, whereas benthic invertebrates at high latitudes often have larvae produced by viviparity, ovoviviparity, and brooding, counteracts Rapoport effects at least in the oceans (see Rohde 1989, 1992 for a discussion and references).

Latitudinal gradients in niche width of parasites of marine fish were examined by Rohde (review 1989, references therein). In the ectoparasitic Monogenea infecting the gills of fish, host ranges, the number of host species used by a parasite species, are more or less the same at all latitudes, although relative species diversity (the number of parasite species per host species) is much greater in tropical waters. In the endoparasitic Digenea, host ranges are larger at high latitudes. However, if the intensity and frequency of infection with particular parasite species is considered (using Rohde's 1980e host specificity index), host specificity is very great and similar at all latitudes, both for the Digenea and Monogenea (Rohde 1978c, 1993). The reason is that, in cold seas, even digenean species with broad host ranges infect very few host species heavily. All this means that niche width, at least along the niche dimension host specificity, is not narrower in high diversity habitats. Microhabitat width of ectoparasites of marine fishes is also not wider in the tropics than in cold-temperate waters.

Alternative explanations

Many authors have used positive correlations to find "explanations" of patterns, ignoring alternative explanations and ignoring historical events. For example, correlations between tree diversity and contemporary climate and, in particular, energy, have been used to claim that explanations of tree diversity based on extant factors are sufficient and that historical processes or events are superfluous towards explaining the patterns (references in McGlone 1996). However, it is important to point out that correlations cannot give an explanation of a pattern, but, at best, suggest an explanation. This point was well made by McGlone (1996) who stated that diversity–energy correlations of tree diversity are strong only at regional scales, and cannot predict diversity at small plots within latitudinal bands, or between continents. Also, "tree diversity cannot have responded to global glacial–interglacial energy fluctuations because plant species cannot evolve that rapidly nor, in most areas of the world, can migration plausibly adjust floral diversity. Thus, contemporary climate or energy, while yielding excellent correlations with plant diversity, has no explanatory power." Contemporary climate merely acts as a surrogate for past climatic changes (see also Latham and Ricklefs 1993; Francis and Currie 1998). More generally, positive correlations between energy input (temperature) and diversity do not explain diversity patterns in the sense that temperature somehow limits diversity (see for example, the recent study by Hawkins *et al.* 2003, who reviewed the empirical literature and concluded that contemporary climate constraints and especially measures of energy, water, or water-energy balance are not the only, but they are the most important factors explaining spatial variation in species richness). Such correlations may also mean that temperature is involved in determining diversity via acceleration of evolutionary processes.

In the following, I discuss such an alternative explanation for latitudinal diversity gradients.

Effective evolutionary time in nonequilibrium ecosystems

Considering the objections to the explanations discussed above and some other explanations, Rohde (1978a,b, 1992, 1999) concluded that none of the explanations for latitudinal gradients in species diversity that assume saturation of habitats with species and equilibrium conditions, and higher "ceilings" of diversity in the tropics due to some limiting factor, is acceptable. Hypotheses based solely on evolutionary or ecological time do not give a satisfactory explanation either, because tropical habitats are not generally older than cold-temperate ones. Latitudinal diversity gradients have been in existence for at least 270 million years (Stehli *et al.* 1969) or may be time-invariant altogether (Crame 2001); there were marked temperature changes in tropical seas in the geologic past, possibly greater than in cold-temperate waters; and mass extinctions have occurred in tropical and cold-water environments (for discussion and references see Rohde 1992). Rohde suggested that available data are best explained by the assumptions that species saturation has not been reached, that higher energy input accelerates evolutionary speed by shortening generation times, increasing mutation rates, and speeding up selection, and that evolutionary speed and the time under which communities have existed under relatively constant conditions (both determining the "effective evolutionary time") are responsible for the gradients. In other words, more species have accumulated in the tropics because evolution there is faster. Essential evidence for this hypothesis is as follows:

(1) diversity has increased over evolutionary time;
(2) extant niche space is not filled, and an increase in diversity is likely (i.e., nonequilibrium conditions prevail);
(3) higher temperature increases speciation rates by (a) shortening generation times, (b) increasing mutation rates, and (c) speeding up selection due to generally accelerated physiological processes (Q_{10}).

According to Jablonski (1999, see also Benton 1995, 1998; Courtillot and Gaudemer 1996; and Jackson and Johnson 2001), there was a sharp rise in diversity in the Cambrian, followed by a Paleozoic plateau interrupted by several mass extinctions, and a sharp rise since the Triassic, also interrupted by several extinction events. Thus, the fossil findings lend strong support to the view that saturation, overall, has not been reached; (see Chapter 6).

Nonsaturation of habitats with species, i.e., the existence of vacant niches, was discussed in detail on pp. 39-48. Here, I repeat the main points. There is indeed ample evidence that many vacant niches exist. This can be shown by comparing species numbers in similar habitats. For example, examinations of many species of marine fish have shown remarkable differences in numbers of parasite species (Figure 2.6). Rohde (1998a) estimated that a maximum of only about 16% of all niches available to metazoan ectoparasites of marine fishes are filled, if the assumption is made that the maximum reached by some fish species is indeed the maximum possible, for which – however – there is no evidence. Walker and Valentine (1984) estimated that 12–54% of niches for marine invertebrates are vacant, and Lawton and collaborators (e.g., Lawton 1984b) demonstrated the availability of many empty niches for insects of bracken (p. 130).

Several studies deal with the effects of temperature on generation times and the effect of generation time on speed of evolution (for an early discussion see Rensch 1954). Tyler *et al.* (1994) have shown that butterflies have shorter generation times in the tropics (for poikilotherms and homoiotherms in general see discussion in Rohde 1992 and references therein). Although there is no evidence for shorter generation times of tropical birds (Mayr 1976), speed of selection (independent of generation time) and mutation rates may still be affected by temperature, and, furthermore, the evolutionary speed of birds may be determined by diversification of taxa with shorter generation times lower down in the ecological hierarchy. Sibley and Ahlquist (1990, further references therein) gave a detailed discussion of the effects of generation time on the rate of DNA evolution, and concluded that "generation time does have an effect on the average rate of genome evolution" (see also Martin and Palumbi 1993, further references in Cardillo 1999). Bromham *et al.* (1996) showed, using 61 mammal species and controlling for phylogeny, that substitution rates at fourfold degenerate sites in two of three protein sequences were negatively correlated with generation time. They found no evidence for any effect due to metabolic rates on sequence evolution (see also Mooers and Harvey 1994). Barraclough *et al.* (1996) found a positive correlation between the rate of evolution of the *rbc*L chloroplast gene with the number of species per family within families of angiosperms, unrelated to generation time. Rosenheim and Tabashnik (1993) evaluated a large data base on the evolution of pesticide resistance of insects and found that the number of generations

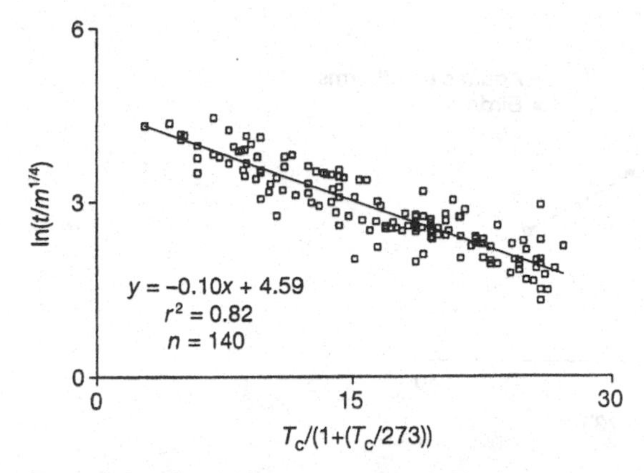

Figure 9.11. Shorter generation times at high temperatures have been suggested to be one important direct cause of latitudinal gradients in species richness. That generation times are indeed negatively correlated with temperature was shown by Gillooly and collaborators for various groups of vertebrates, insects, and zooplankton. The example illustrated in this figure shows the effect of incubation temperature on mass-corrected embryonic development time for marine fishes in the field. Other examples are illustrated in Figure 9.12. Reprinted from Gillooly, Charnov, West, Savage, Van, and Brown (2002), with the permission of MacMillan Publishing Ltd., and the authors.

per year made no contribution to the ability to evolve resistance. However, a more complete model, also using the effects of pest severity and pest feeding mode, showed a significant curvilinear effect for generations per year: species with intermediate numbers of generations per year were best able to evolve resistance.

Recently, Gillooly *et al.* (2002) described a general model, based on first principles of allometry and biochemical kinetics, that predicts generation time ("time of ontogenetic development") as a function of body mass and temperature. The model fits data for embryonic development times from birds, fish, amphibians, aquatic insects, and zooplankton, and also describes development of other animals at other life-cycle stages. Invariably, development time of species in all of these groups is negatively correlated with temperature (Figures 9.11, 9.12).

The fossil record shows that some invertebrates with short generation times, e.g., molluscs, have hardly changed over millions of years, whereas vertebrates with relatively long generation times have undergone remarkable changes over much shorter periods (references in Rohde 1992). But this observation does not affect our considerations of latitudinal diversity gradients. We do not compare species richness of molluscs at one latitude

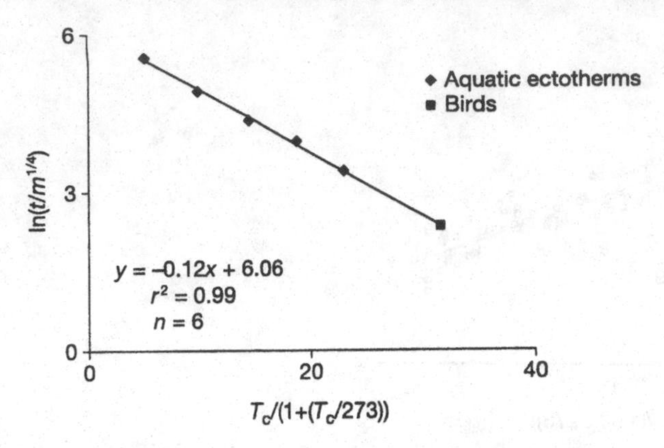

Figure 9.12. Effect of incubation temperature on mass-corrected embryonic development time for aquatic ectotherms (fish, amphibians, zooplankton, aquatic insects) and birds. Reprinted from Gillooly, Charnov, West, Savage, Van, and Brown (2002), with the permission of MacMillan Publishing Ltd., and the authors.

with that of vertebrates at another, i.e., comparisons should not be made across phyla, but should be restricted to lower taxa.

We conclude that, although evidence is still limited, the important study of Gillooly *et al.* (2002), in particular, demonstrates that generally an increase in temperature does indeed shorten generation times, supporting one key element in the hypothesis that effective evolutionary time determines latitudinal gradients in diversity.

Early studies of the effect of temperature on mutation rates were discussed by Rohde (1992, references therein). There is indeed evidence that raised temperature increases mutation rates (e.g., Timoféeff-Ressovsky *et al.* 1935; Bazin *et al.* 1997), but more studies are needed to provide evidence for more taxa and a variety of genes.

There is a wealth of evidence showing an acceleration of physiological processes with temperature. Pandian and Vivekanandan (1985) have shown that maintenance metabolism, food intake, and growth are higher in tropical than in cold-water fish, and according to Tarr (1969), high temperatures lead to a decrease in metabolic loss and an increase in food consumption of fish. Although most selection is characterized by stasis, and although mutations and selection do not determine speed of speciation on their own, but in conjunction with mechanisms that bring about genetic isolation, it seems intuitively likely that, under otherwise identical conditions, species that have higher mutation rates, faster physiological processes, and shorter generation times, will more frequently encounter conditions where genetic isolation can occur. One can model this by using "pitted speciation landscapes" (compare the "rugged fitness landscapes" used in population genetics): such a landscape has many "pits" which are regions where isolation is achieved. Selection will normally try to keep species in these pits (i.e., there is stasis), and the pits are of course not permanent but evolve depending on various changing factors, such as the presence of other species. Overall, populations that change faster, driven by accelerated mutation rates, faster selection, and shorter generation times, will move around the speciation landscape faster and "fall" more frequently into the isolation pits. However, experimental evidence for direct temperature effects on speed of election are not available. Such evidence is urgently needed.

The combined effects of the above should be an increase in speciation rates at higher temperatures. Rohde (1992) gives some references in support of this assumption. Jablonski (1993) presented paleontological evidence that post-Paleozoic marine orders have appeared more frequently in tropical waters. Cardillo (1999) found, for birds and butterflies, that relative rates of diversification per unit time increase towards the tropics, (but see Bromham and Cardillo 2003, who found no effect of latitude on evolutionary rates in 45 pairs of phylogenetically independent bird species using two mitochondrial genes and DNA–DNA hybridization distances). Most importantly, Allen *et al.* (2002) have shown that species diversity can be predicted from the biochemical kinetics of metabolism. Their model predicts quantitatively how species richness increases with environmental temperature. They conclude that "evolutionary rates are ultimately constrained by generation times of individuals and mutation rates. Both of these rates are correlated with metabolic rates and show the same Boltzmann relation to temperature. The results therefore support the hypothesis that elevated temperatures increase the standing stock of species by accelerating the biochemical reactions that control speciation rates." (Figures 9.13, 9.14). Harmelin-Vivien's (2002) conclusion of her chapter on "Energetics and Fish Diversity on Coral Reefs" is very similar to the hypothesis that effective evolutionary time is largely responsible for the latitudinal gradients in species diversity. According to her, synergy between constant high temperature and long-term stability may speed up the rate of molecular evolution and increase genome variability allowing the emergence of new species by natural selection. Most recently, Wright *et al.* (2003) provided support for the hypothesis that speciation rates are higher in

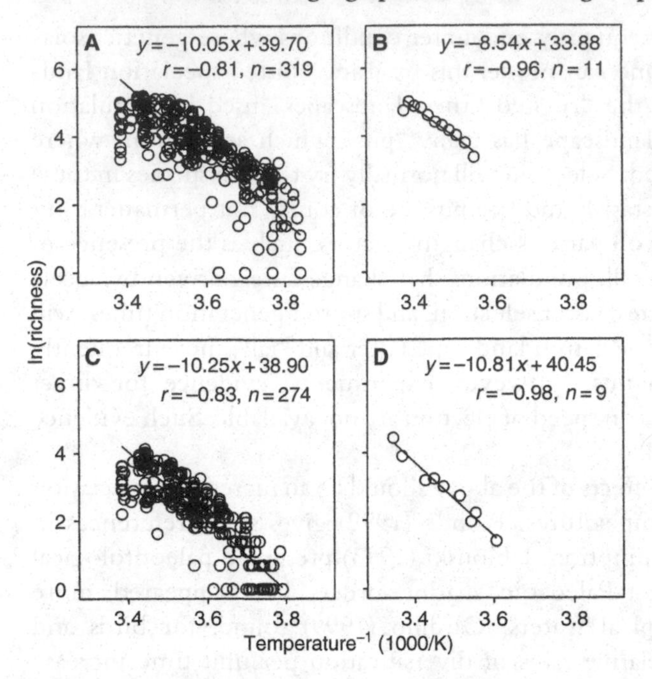

Figure 9.13. The prediction that direct temperature effects on speciation rates are responsible for latitudinal gradients in species diversity was tested by Allen and collaborators, who have shown that diversity can be predicted from the biochemical kinetics of metabolism, as shown by the examples which illustrate the effects of mean ambient temperatures on species richness of North American trees (A), Costa Rican trees along a 2600 m elevational gradient on Volcan Barva (B), North American amphibians (C), and Ecuadorian amphibians along a 4000 m elevational gradient in the Andes (D). From Allen, Brown, and Gillooly (2002). Reprinted by permission of the authors and the American Association for the Advancement of Science.

the tropics. They compared rDNA substitution rates for a group of closely related plant species from different biomes in the western Pacific. Rates were indeed higher in habitats with greater biologically available energy. Martin and McKay (2004) compared the within-species patterns of mitochondrial DNA variation across 60 vertebrate species from two oceans, six continents, and 119 degrees latitude. After controlling for geographic distance, they found greater genetic divergence of populations within species at lower latitudes, providing further strong support for the hypothesis. Kaspari *et al.* (2004) have shown that the energy-speciation hypothesis is the best predictor of ant species richness, i.e., it predicts the slope of the temperature diversity curve and accounts for most of the variation in diversity.

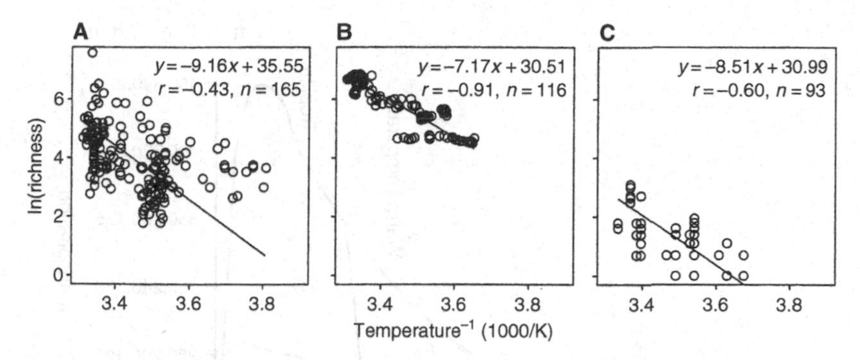

Figure 9.14. Effects of mean annual water temperature on species richness of fish (A), numbers of marine prosobranch snails per latitudinal degree band along the continental shelves of North and South America (B), and ectoparasite species per host of marine teleost fish from Antarctica to the tropics (C). From Allen, Brown, and Gillooly (2002). Reprinted by permission of the authors and the American Association for the Advancement of Science.

In summary, all the latitudinal differences supposedly "explaining" gradients in species diversity by assuming that communities are saturated and in equilibrium can also be the result of and are indeed better explained by a gradient in "effective evolutionary time", i.e., in speed of evolution and length of geological time over which communities have existed under more or less the same conditions. However, of course this does not imply that other factors, such as area, heterogeneity, and local history, may not play some role as well.

General global patterns in diversity

Latitudinal gradients in diversity are the most conspicuous and the best studied diversity gradients, but other gradients exist. The overall aim should be to establish models that describe global patterns in diversity incorporating latitude, longitude, differences in rainfall, etc. Some recent studies attempt to do this (e.g., Ricklefs 2004). At least some of these models can be reconciled with the hypothesis of effective evolutionary time, even if they do not specifically consider this. Thus, Zhang and Wu (2002) developed a statistical thermodynamic model of the organizational order of vegetation (OOV) that can be used to derive vegetation patterns on a large scale. OOV is defined as "a thermodynamic measure of the degree of structural and functional self-organization of natural vegetation in a given environment,

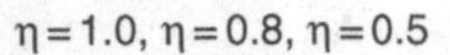

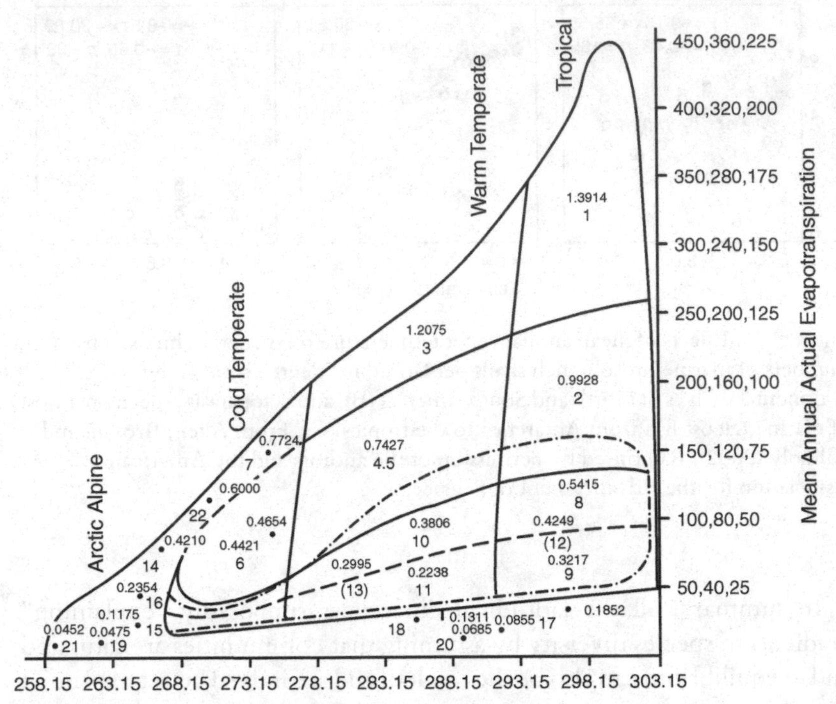

Figure 9.15. Organizational order of vegetation (OOV) of the major world biomes as classified by Whittaker (1975). 1 – tropical rainforest, 2 – tropical seasonal forest, 3 – temperate rainforest, 4 – temperate deciduous forest, 5 – temperate evergreen forest, 6 – taiga, 7 – elfinwood, 8 – tropical broadleaf woodland, 9 – thorn scrub, 10 – temperate woodland, 11 – temperate shrubland, 12 – savanna, 13 – grassland, 14 – alpine shrubland, 15 – alpine grassland, 16 – tundra, 17 – warm semidesert, 18 – cool semidesert, 19 – arctic alpine semidesert, 20 – true desert, 21 – arctic alpine desert, 22 – cool temperate bog. Small numbers with decimal points are the average values of OOV for the 22 biomes when $\eta = 0.8$. For biomes without closed boundaries, the value of the marked point for each type is given. The actual OOV values are those in the figures times 10^{33}. Note that species richness is correlated with temperature and mean annual precipitation. From Zhang and Wu (2002). Reprinted with permission of Elsevier and the authors.

which is related to the complexity and stability of ecosystems". The model derives the maximum possible ecoclines and predicts smallest OOV values in cold-dry environments (Arctic, Alpine), and highest values for moist warm environments (wet tropics) (Figure 9.15). Criddle *et al.* (2003)

developed the hypothesis that the primary climatic variables responsible for global-scale gradients in diversity and ranges of ectotherms are temperature and temperature variability. Respiratory energy metabolism is the primary physiological variable that defines adaptations of ectotherms to temperature. Therefore, adaptations to latitudinal and altitudinal gradients of temperature and temperature variability lead to corresponding gradients in properties of energy metabolism, which give rise to gradients in species diversity and ranges.

10 · *An autecological comparison: the ecology of some Aspidogastrea*

In the previous chapters, we have evaluated evidence for equilibrium and nonequilibrium conditions, and for the significance of interspecific competition, using ecological and macroecological/biogeographical, as well as some experimental studies. Here we look at two closely related species using everything that is known about them, including phylogeny based on DNA and morphology (including ultrastructure), life cycles, morphology and ecological evidence, in order to arrive at an understanding of whether and to what degree competition can be implicated in the evolution of the species' characteristics, i.e., we use an autecological approach.

Hengeveld and Walter in a number of papers (e.g., 1999), advocated a radically new approach to ecology (see pp. 10–11). They distinguish two coexisting but mutually exclusive paradigms, the demographic and the autecological paradigm. Autecology, the study of the ecology of a single species, its phylogenetic history and adaptations, is an old concept, new is the radical view that these two approaches are incompatible. According to Hengeveld and Walter, the demographic approach is based on unrealistic premises and cannot answer many of the questions posed by the way species are distributed and how common they are.

The examples selected are two species of the Aspidogastrea. I chose this group for various reasons. Firstly, I am very familiar with it, having done a lot of work on its phylogeny based on molecular, morphological, and ultrastructural data, as well as on the taxonomy, development, and ecology of several species. Secondly, the group is better known than many others, although much remains to be done particularly on the physiology and ecology of many species. Thirdly, the group is small, comprising only 12 genera (in 4 families) and about 80 described species, and its relative uniformity permits insights into some essential aspects of its ecology. A "Tree of Life" webpage at http://tolweb.org/tree/eukaryotes/animals/platyhelminthes/aspidogastrea/ contains a complete list of references and illustrated discussions of the morphology, taxonomy, pathology, ultrastructure, life cycles, and embryonic development of the group, to which the reader is referred. Illustrations for the discussion presented here can therefore be kept to a minimum.

The Aspidogastrea (=Aspidobothrea=Aspidobothria) is a taxon belonging to the phylum Platyhelminthes. Within the phylum, it is the sister group to the digenean trematodes, as shown unambiguously by morphology, ultrastructure, and molecular data. It contains species parasitic in molluscs and vertebrates. For some species, molluscs serve as intermediate hosts harboring larvae and juveniles, for others, they may also act as definitive hosts in which maturation of the worms occurs. Vertebrates become infected by eating infected molluscs, at least in the species for which the life cycle is known. Aspidogastreans are likely to be very ancient, originating probably several hundred million years ago (Figure 10.1). There is controversy about whether molluscs or vertebrates are the original hosts. Three of the four families, each with a single genus and one or two species, infect chondrichthyan fishes (sharks, rays, and chimaeras), species of the fourth family occur in teleost fishes and turtles. Chondrichthyans are over 400 million years old, and the few digenean trematodes infecting them have almost certainly been acquired secondarily from teleost fishes. On the other hand, the relatively great diversity of aspidogastreans in chondrichthyans at least at the family and genus level suggests that these fish are their original hosts.

Aspidogastreans are organisms of remarkable complexity. Whereas almost all digenean trematodes possess one or two suckers, the aspidogastreans have a row of ventral suckers or a ventral disc that is subdivided into many suckerlets. The nervous system of the only species, *Multicotyle purvisi*, examined in detail to date has more longitudinal and transverse nerves than any other platyhelminth, and there is an extraordinary variety of sensory receptors. Aspidogastreans appear to be only "superficially" adapted to a parasitic way of life. Whereas digenean trematodes can be kept alive outside a host only in artifical media of great complexity, simulating the environment of the host and the chemical compounds provided by it, aspidogastreans have been kept alive outside a host for many days or even weeks in water or saline solution. Also, host specificity among aspidogastreans is very low, that is, a species typically infects many molluscan and vertebrate hosts, or – if a species is specific – specificity seems to be due to ecological and not physiological factors. However, many species have not been examined in this way.

In the following, I discuss two species in detail, *Multicotyle purvisi* and *Lobatostoma manteri* (Rohde 1968, 1972, further references therein, 1973,

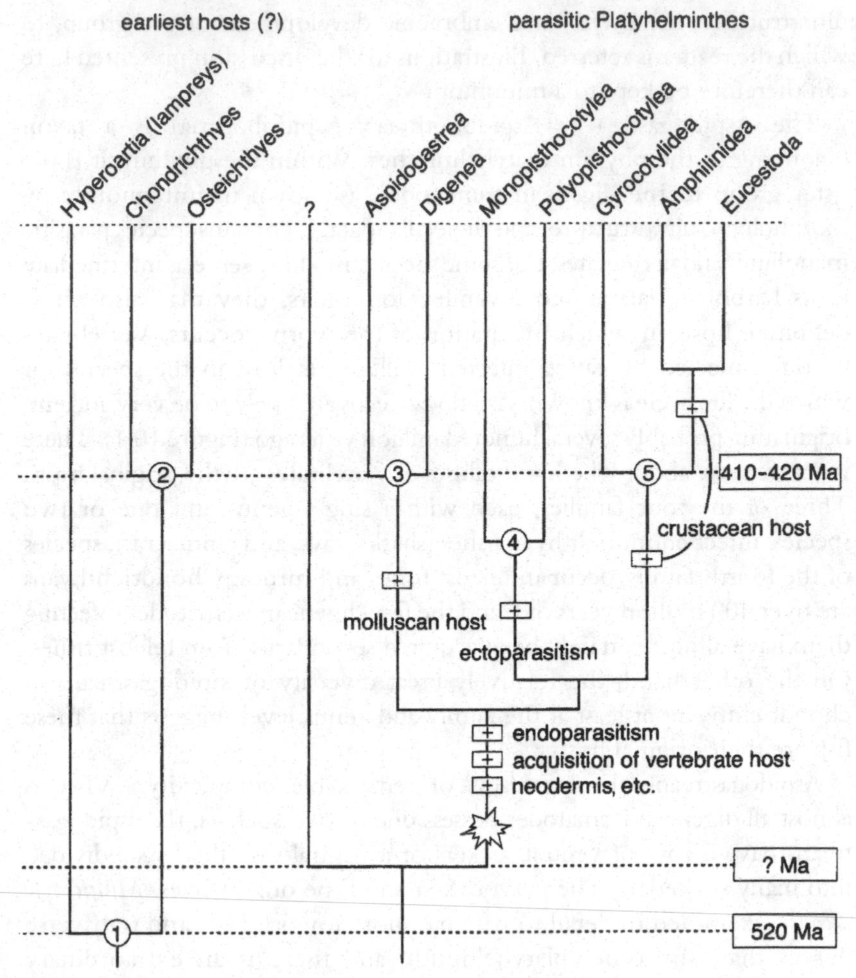

Figure 10.1. Phylogeny of Platyhelminthes according to Littlewood *et al.* (1999), modified from Rohde (2001b). Note that the Aspidogastrea diverged from its sister group, the Digenea, about 410–420 million years ago. From Rohde (2001b). Reprinted by permission of Taylor & Francis.

1975, 1981b, 1994b, further references therein, 2001b; Littlewood *et al.* 1999). Both species belong to the family Aspidogastridae and use snails as intermediate hosts. The former species uses freshwater turtles and the latter marine fish as final hosts.

Multicotyle purvisi

Multicotyle purvisi naturally infects the stomach and duodenum of several freshwater turtles in SE Asia. Ectolecithal eggs are laid into freshwater at the 1–3 cell stage. At 27–29 °C mature larvae develop in about 25 days. Hatching usually occurs in the morning, released by light stimulus, but also occurs without such a stimulus when permanently kept in the dark. Unique in the animal kingdom, the larva (about 200 μm long) is covered by a tegument drawn out into numerous very thin (120–180 Å diameter) processes, the so-called microfila (Figure 10.2). The microfila are longest at the anterior end, where they reach a length of about 6 μm. The larva has a large posterior sucker as well as two dorsal, two ventral, and six posterior ciliary tufts responsible for the swimming movements of the larva. After hatching, the larva creeps on the ground or swims freely in the water. It floats for extended periods in the water, apparently assisted by the microfila which enlarge the surface area and prevent the larva from rapidly sinking to the bottom. A larva often remains attached to the water surface membrane, before detaching and sinking slowly downwards or remaining in the water column. It survives for maximally about 33 hours in the water. While floating in the water column, it is carried by inhalation currents into the breathing chamber of a snail, from where it migrates into the snail's kidneys, where it grows up. Snails of three families, Viviparidae, Ampullariidae, and Bithyniidae, could be infected experimentally. Speed of development within the snail host depends on the snail species and temperature. Worms from snails fed to turtles led to successful infection.

Larvae possess an extraordinary array of sensory receptors, which have been examined in detail by electron microscopy. They include a pair of eyes, a pair of receptor complexes near the anterior end, each consisting of a cavity containing one normal and several modified dendritic sensory nerve endings, and 11 unciliated or uniciliated receptor types. The functions of the various types of receptors are unknown, but it is likely that the eyes contribute to hatching and possibly to swimming upwards in the water column. The other receptors may help in migration to the snail's kidneys.

Turtles become infected by eating infected snails. Adult worms reach a length of about 10 mm or more. Like the larvae, they have a great variety of sensory receptors, 7 or even 9 types, and a remarkable nervous system: anteriorly there are not – as is usual in flatworms – a number of single circular commissures, but rather of two, internal and external ones, one of the former very large and acting as a cerebral commissure. The function of receptors and the nervous system is not known, but they must contribute

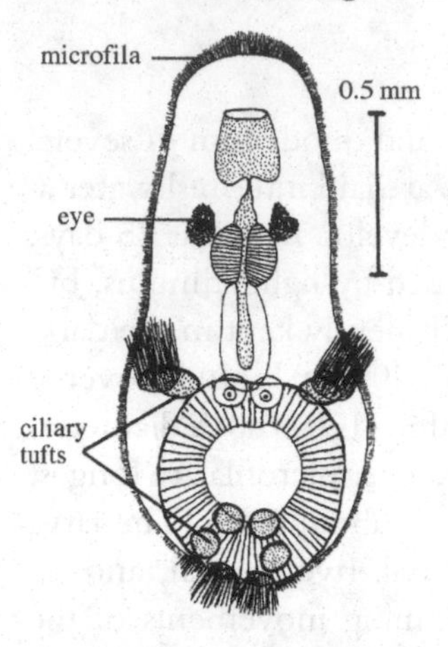

Figure 10.2. Larva of *Multicotyle purvisi* (Aspidogastrea). Note the thick layer of microfila, the ciliary tufts (two ventral, two dorsal, six posterior), the two large eyes, the large posterior sucker, and the lack of an anterior sucker. Redrawn and modified from Rohde (1968).

to niche finding in the host (i.e., migration to the stomach/duodenum), feeding, or mate finding. Alternatively, or additionally, they may help in preventing damage by the juvenile parasite to the delicate snail-host tissue.

The ultimate aims of autecology are to elucidate environmental conditions that, in this case, lead to infection of intermediate and final hosts, and to the persistence of populations and species. We know that flotation of larvae in the water column leads to infection of snails, future studies should try to quantify the conditions that are the prerequisites for infection. How strong must the inhalation currents of snails be to guarantee infection? Are small snails that have weak inhalation currents unsuitable hosts? What are the effects of water temperature and strength of water currents on the likelihood of infection, etc.? Why are some snails and turtles of one particular species infected while others are not? Likely candidates to explain the differences are the conditions of the abiotic environment that determine whether snails can exist at sufficient densities to provide minimum population densities of parasites necessary to keep the life cycle in a certain habitat going.

Lobatostoma manteri

Adult worms of this species (maximum length about 5 mm) live in the small intestine of a marine teleost fish, the snub-nosed dart *Trachinotus*

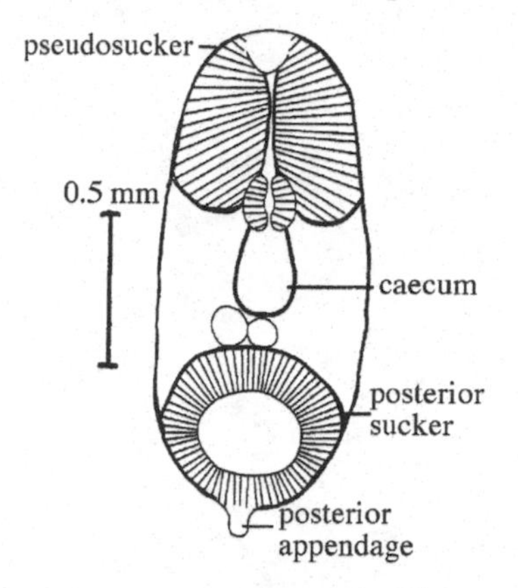

Figure 10.3. Larva of *Lobatostoma manteri*. Note the large posterior sucker and the muscular anterior pseudosucker, lack of microfila, ciliary tufts, and eyes. Redrawn and modified from Rohde (1973).

blochi, on the Great Barrier Reef, Australia. Eggs containing fully developed larvae are laid. Larvae are about 150–200 μm long and lack the layer of microfila found in *Multicotyle* (Figure 10.3). They have an anterior so-called pseudosucker and a posterior sucker, but lack ciliary tufts. They have nine receptor types, but no eyes or sensory complex. Eggs are eaten by snails, and larvae hatch in their stomach. Three snail species, *Cerithium (Clypeomorus) moniliferum*, *Planaxis sulcatus*, and *Peristernia australiensis*, all of which live on the beachrock of coral cays, were found to be naturally infected at Heron Island, Great Barrier Reef. The first two species were also used in infection experiments, with positive results. Larvae migrate immediately after hatching along the ducts of the digestive gland into its follicles. Larvae feed on the secretion and probably epithelial cells of the digestive gland. The posterior sucker is used for adhesion to the epithelium and contributes to its erosion. Some young worms were also found in the stomach of *Planaxis*, and up to six worms in the stomach and large digestive ducts of *Peristernia*. These two species are larger than *Cerithium*. Infective juveniles in snails are almost as large as adults in fish and even contain well developed genitalia and young sperm and egg cells, which – however – do not mature. Fish become infected by eating snails, as also demonstrated by infection experiments. Snails are very thick-shelled and must be crushed between the pharyngeal plates of the snub-nosed dart. Fish of similar size lacking such strongly developed pharyngeal plates cannot become infected.

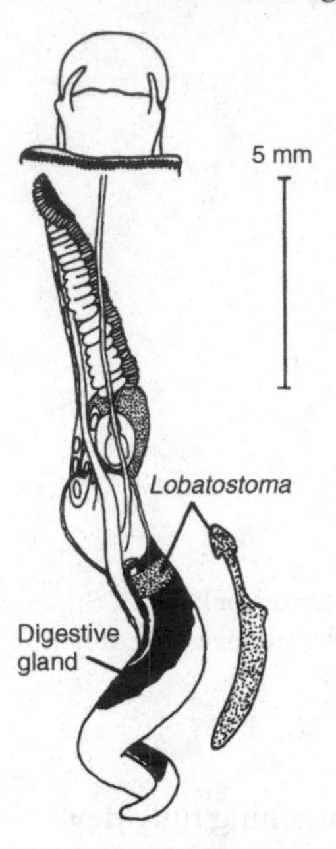

Figure 10.4. A juvenile *Lobatostoma manteri* (dotted) in a cavity of the digestive gland of the snail *Cerithium moniliferum*. The same extended worm, dissected out of the snail, to the right of the snail. Note the large size of the worm relative to that of the snail. Redrawn and modified from Rohde (1973).

Worms from fish can be kept alive in dilute seawater for up to 13 days, and they continue to produce eggs infective to snails. Juveniles from snails stay alive in seawater and dilute seawater.

Electron microscopic investigations have shown that juvenile and adult worms have 8 and possibly 14 types of sensory receptors, which differ in the presence or absence of cilia, in the shape and length of cilia, and in the presence or absence of ciliary rootlets. Juvenile *Lobatostoma* from snails, on the basis of serial sections and scanning electron microscopy, were estimated to have at least 8500 surface and numerous sub-surface receptors. The function of the receptors is not known, but they are likely to play a role in migration from the stomach to the small intestine, feeding, and mate finding, and/or they may help in preventing damage to snail-host tissue.

There is evidence for intraspecific competition leading to density dependence in *Lobatostoma*. The smallest of the three snail species found to be infected, *Cerithium moniliferum*, almost always harbors only a single individual parasite, coiled up in a cavity formed by the main and one (or some?) side ducts of the digestive gland (Figure 10.4), whereas *Peristernia australiensis* harbors up to six in the stomach and large digestive ducts. There is simply not sufficient space for larger numbers of parasites in *Cerithium* (as well as *Peristernia*). In other words, it is likely that intraspecific competition for space and food limits parasite numbers. Both *Cerithium* and *Peristernia* are also hosts to several species of digenean trematodes, some occurring at high prevalences of infection. The population dynamics of snails and parasites were studied over several years at Heron Island, Great Barrier Reef. Over a period of presumably unfavourable conditions between January 1971 and April 1972, there was a strong decrease in the relative numbers of infected *Cerithium* and *Peristernia*. Snails with double infections (Digenea plus *Lobatostoma*) disappeared first, i.e., the co-occurrence of other parasite species affected the survival chances of the aspidogastrean. Infection with *Lobatostoma* did not affect the relative numbers of egg-producing snails during the period of high prevalence of infection.

Snails that are suitable hosts were spread all around the Heron Island, but only one small flat-bottomed area, "Shark Bay", had snail populations with a high prevalence of infection. The reason is as follows: schools of juvenile dart follow the incoming tide and feed on large numbers of snails, indicated by the many shell fragments scattered over the sea bottom at Shark Bay. At the same time, worm eggs are shed which are eaten by the snails. In other regions, many of which are inhabited by the same snail species, few snails are infected or none at all, because the sea floor is too uneven or too sandy, or the wave action is too strong, for any significant level of infection of snails to occur. In other words, the local distribution of the parasite is not determined by resource availability, but is the consequence of interactions between the physical environment, the hosts and the infective stages of the parasite. Future research should attempt, in line with the autecological paradigm, to quantify these interactions. Which abiotic factors determine the distribution of the snail and fish hosts, under which conditions can infective stages of the parasites within the eggs survive long enough to facilitate infection of the snails, and how do movements of snails in different localities affect the likelihood of being eaten by the fish?

Comparison of the species, and coevolution

Looking back at and comparing the two species, we find distinct differences. Each is adapted to its particular host and has evolved morphological adaptations and life cycles ensuring transmission. In *Lobatostoma manteri*, eggs are laid containing fully developed larvae that are eaten by snails and hatch immediately. If eggs were laid at an earlier stage of development, the chances of infecting snails would probably be reduced to zero, because eggs would not survive in the snail if eaten immediately, and they would not survive on the sea floor long enough to complete development, due to periodic exposure at low tides. There is a relatively small number of sensory receptors, because an intermediate host is not actively searched for. In *Multicotyle purvisi*, eggs are laid that have to develop over almost a month on the bottom of freshwater bodies before they can hatch. This facilitates production of larger numbers of eggs than if eggs containing fully developed larvae were produced, increasing chances of infection. On the other hand, it restricts completion of the life cycle to calm waters, because in strong currents eggs as well as larvae floating in the water would be swept away. Hatching occurs in the morning (probably facilitated by the eyes) to ensure that snails, which are apparently active during the day, become infected. Larvae have an astonishing array of sensory receptors, ciliary tufts, and a dense layer of microfila, to ensure swimming and floating in the water column and inhalation by a snail host. Larvae, furthermore, have to find their way to the kidneys, probably also facilitated by the receptors. The very great number of receptors in juveniles and adults is likely to play a role inside the vertebrate host. Mate finding, feeding, and (not very likely, because the stomach and small intestine can hardly be missed) finding the way to their microhabitat are the only possibilities. They may also help in reducing damage to the delicate snail tissue. Lack of an anterior "pseudosucker" in *Multicotyle* may be an adaptation to the delicate kidney tissues, in which the juvenile develops. Damage to the host may thus be avoided. In contrast, the large and muscular pseudosucker of *Lobatostoma* may help in feeding, i.e., biting off pieces of robust host tissue, the digestive gland of snails.

As just mentioned, intermediate hosts are also infected with digenean trematodes, but there is no evidence whatsoever that interspecific competition has been in any way involved in the evolution of the very intricate morphological and behavioral adaptations of the two aspidogastrean species. In other words, there is no evidence for coevolution with other, competing species. This is also the case if we consider other aspidogastreans. *Rugogaster* infects the caecal glands of chimaeras, *Multicalyx* the gall bladder and bile ducts of chimaeras, sharks, and rays, and *Stichocotyle* the bile ducts of rays. Each species has adapted to its host and microhabitat within the host over millions of years, although we do not know what these adaptations are. There are many more potential host species than aspidogastrean species infecting them and their particular microhabitats. And there are no potentially competing species of other taxa in these microhabitats. In other words, there is a vast number of vacant niches. It seems, then, that at least for the aspidogastreans, the autecological paradigm is more useful than the demographic one. Demographic ecologists have not found the aspidogastreans worthy of their attention because they are rare and unlikely to compete with each other or, to any significant extent, with species of other taxa.

Generality of the approach

The question of whether the aspidogastreans are exceptional must be answered, and also to what extent is the autecological approach useful? Aspidogastreans are very ancient and can be considered as living fossils. They may have lost the versatility of undergoing much further adaptive radiation, or – in the terminology of Kauffman (1993) – they may have reached radiation and fitness stasis as a consequence of adaptive walks in rugged fitness landscapes which have led them to local optima where they are trapped. Species that have not had such a long evolutionary history, perhaps many insects, may still be relatively unfit and still climbing in many directions, where the chances to encounter other, potentially competing, species still exist. So, the possibility should be considered that the autecological approach is best suited for slowly evolving taxa with little adaptive potential and little speciation, whereas the demographic approach is better suited to rapidly evolving taxa which can adapt rapidly and speciate fast.

11 · *What explains the differences found? A summary, and prospects for an ecology of the future*

What explains the differences between communities?

Rohde (1980a) suggested that animal communities can be arranged in a continuum from random and unstructured to highly structured, depending on ecological characteristics of species in the communities. Animals with little vagility and/or small population or individual size live in largely empty niche space. They are less subject to structuring mechanisms, in particular competition, than are large animals or animals that live in large populations with much vagility (although they may be nonrandom to a degree because of nonrandom colonization events). The latter have filled extant niche space to a greater degree, i.e., they are closer to saturation, and include the predominantly large mammals and birds, and free-living vagile insects occurring in large populations. (Saturation, however, does not exclude the possibility of further increases in diversity by subdivision of niches.) Gotelli and Rohde (2002) tested this hypothesis using null-model analysis to check for nonrandomness in the structure of metazoan ectoparasites of 45 species of marine fish, and compared the results with those for herps, birds, and mammals. In parasites, co-occurrence patterns could not be distinguished from those that might arise by random colonization and extinction. Presence–absence matrices for small-bodied taxa (parasites, herps) with low vagility and/or small population size were mostly random, whereas presence–absence matrices for large-bodied taxa with high vagility and/or large population size (birds, mammals) were highly structured, supporting Rohde's hypothesis. For Figure 11.1, some data from Gotelli and McCabe (2002) were also used, which support the hypothesis even more strongly. It must be realized, however, that the null-model analysis, even if it reveals structure in communities, says nothing about the mechanisms responsible for structuring: interspecific competition – suggested to be responsible by Diamond (1975) – is

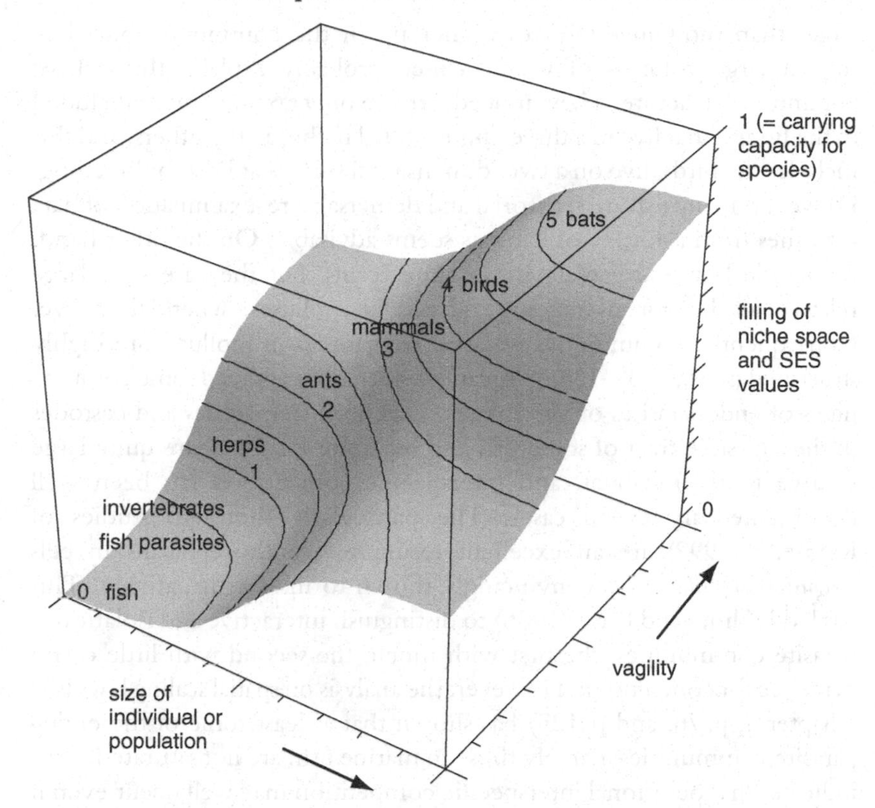

Figure 11.1. Diagram representing filling of extant niche space as indicated by SES values (standardized effect sizes, see Gotelli and Rohde 2002, and Figure 8.4) for various vertebrate and invertebrate groups. Note that taxa comprising large (relative to their habitat) species and/or species with great vagility (bats, birds, mammals, ants) have greater SES values than taxa containing small species and/or species with little vagility. Data from Gotelli and Rohde (2002) and Gotelli and McCabe (2002).

only one of many mechanisms. Others include the affinities of species for non-overlapping habitats leading to segregation, events that reflect biogeographic and evolutionary history, etc. (references in Gotelli and McCabe 2002).

In interpreting Figure 11.1, two important points must be kept in mind: (1) vagility and/or dispersal may be more important than body size, because populations of animals at the top of the food chain often have smaller biomass than smaller herbivores "below" them; (2) the body size of species has to be measured relative to their habitat. Thus, fish, which have the lowest standardized effect sizes (SES) values of all communities tested, are larger than most invertebrates higher up in the continuum, but they inhabit large volumes of water, which probably explains the lack of community structure. They, indeed, are the only taxon of those included in the figure, that live in a three-dimensional niche, all the others, and this includes the birds, live on a two-dimensional surface at least for breeding. However, many fish are territorial and demersal: a re-examination of data for fishes from a variety of habitats seems advisable. On the other hand, trematode larvae are small, in absolute terms, but they are very large relative to the gonads/digestive glands of molluscs, where they live. Consequently, communities of larval trematodes in molluscs are highly structured (see pp. 131–134). Similar considerations apply to some communities of endoparasites of vertebrates. Digenean trematodes and cestodes in the digestive tract of some fish and birds, for example, are quite large relative to their habitat, and interspecific competition has been well documented in several cases. The particularly thorough studies of Kennedy (1992) are an excellent example: acanthocephalans in eels *Anguilla anguilla*, were convincingly shown to interact negatively. This has led Holmes and Price (1986) to distinguish interactive and isolationist parasite communities, the first with much, the second with little or no evidence for competition. However, the analysis on spatial scaling laws (see Chapter 4, p. 76, and p. 125) has shown that at least some of the endoparasite communities, namely those in marine fish, are not saturated. This indicates that occasional interspecific competition may well occur even if species are not densely packed, but that interactions have been without evolutionary significance. Indeed, it is not unlikely that weak interactions will be found in all or most communities, if investigators persist and studies are of sufficient depth and duration. Therefore, it is doubtful that the distinction between interactive and isolationist communities is of much use, as already tentatively suggested by Holmes and Price (1986).

Many important studies have used modelling to make predictions, for example on the effects of habitat fragmentation or reductions in species diversity caused by humans. The example of the Glanville fritillary has been discussed above (pp. 20–22). Many of these models use a demographic approach and we have to ask whether they are compatible with the conclusion that many systems are nonsaturated and not in equilibrium. According to Levin (1998), who was one of the early principal promoters of nonequilibrial approaches to community ecology, patterns at higher levels of ecosystems arise from localized interactions and selection processes at lower levels. His statement that "tight linkages between members of interacting species provide reliable and rapid feedbacks for individual behaviour, the essentials for tight coevolution" is certainly correct to an extent, but many (and probably most) species do not have such tight relationships, and coevolution for them is unlikely. Levin's (2000) assumption of the existence of multiple stable states, with the possibility of "flips" from one to another, also applies to closely knit systems, i.e., to systems which are close to saturation, containing at least some species that are common and widespread, have large SES values (Figure 11.1), and much potential for interaction. It is doubtful that they hold for systems consisting of rare species with little vagility and dispersal abilities, such as those discussed in Chapter 10. Indeed, this is recognized by Levin (1998) by distinguishing keystone species and, more importantly, "keystone functional groups", that is, sets of species that control critical ecosystem processes. Therefore, the question of whether some of the components of ecosystems are nonsaturated and nonequilibrial may be, to some degree, irrelevant for the ecosystem. Nevertheless, the discussion in this book has shown that nonsaturated, nonequilibrial systems which are not tightly knit, are the rule rather than the exception, and tests are therefore necessary to determine how the various models are affected by this fact. Also, species not in any tight bondage to the system as a whole may still be affected by the system, but the effects are largely or entirely top-down. Thus, if there is indeed a flip from one stable state to another, it may devastate habitats on a large scale and bring about extinction of many "innocent" species which have contributed nothing to the demise.

Hubbell's (2001) neutral theory of biodiversity and biogeography proceeds from MacArthur and Wilson's equilibrium theory of island biogeography, including a process of speciation and assuming neutrality not for species but for individuals. It is limited to communities at a particular trophic level, and makes predictions about the relative abundance of species, species–area relationships, phylogeny under genetic drift, random dispersal, and random speciation. It predicts species richness not only on islands but on the mainland as well, and derives a fundamental biodiversity number. It claims to reconcile the niche-assembly and dispersal-assembly perspectives of ecological communities, the former equilibrial and the latter nonequilibrial in nature. Furthermore, the theory predicts that phylogenetic clades are fractal and self-similar on all taxonomic scales, implying that biodiversity is fractal. Ritchie and Olff (1999) have developed packing rules based on fractal geometry and have shown that they apply to herbivorous mammals and savanna plants. Rohde (2001a), however, has shown that the rules are not applicable to a large group of parasites and probably not to the vast majority of animal species (pp. 41–44, 76). They are likely to apply only to those species that are indeed closely packed, i.e., use up much of the resources required by them, thereby competing for them. These taxa have large SES values, as illustrated in Figure 11.1. Generalising these findings, it appears that the neutral theory of biodiversity is applicable only to species high in the ecological hierarchy.

A summary, and prospects for an ecology of the future

There is evidence for equilibrium conditions at all levels. Thus, there has been evolutionary stasis over long periods (see, for instance, Ordovician benthos); some communities appear to be saturated and in apparent dynamic equilibrium, at least for certain periods (e.g., island communities of birds); and equilibrium conditions are common in populations (e.g., some insect populations). However, at all of these levels, periods of equilibrium are often interrupted by even longer periods of nonequilibrium. Thus, in evolutionary history, periods of increasing diversity are more pronounced than periods of stasis, and there has been a significant increase in diversity over evolutionary time. Evidence for nonsaturation and nonequilibrium in ecological communities is more convincing than evidence for saturation, as shown, for example, by very large differences in species numbers of parasites infecting different fish species, and by temporary reductions in species richness due to disturbances. At the level of populations, environmental disruptions are prevalent and so common that most populations probably cannot persist in equilibrium for prolonged periods, and re-establishment of equilibrium conditions may take many years, as shown by some well documented examples of insect populations. Nonequilibrium at the different levels is due to different mechanisms: at the population level, nonequilibria are mainly due to environmental (biotic and abiotic) disturbances; at the community level, they are mainly due to the existence of vast numbers of vacant niches but also disturbances; on an evolutionary scale, they are due to mass extinctions, numerous vacant niches in communities, and as yet uncolonized vast habitats, where whole ecosystems can still be established. For example, the muddy ocean floor was first colonized in the early Ordovician leading to rapid diversification of benthic animals, and invasion of land led to an explosion in the numbers of species of various taxa. Diversity was reduced several times as a result of crashes caused by external factors (e.g., mass extinctions due to asteroid impacts or other events), and periods of long evolutionary stasis were apparently due to a loss of evolutionary versatility with time (Kauffman 1993; Rosenzweig 1995), but these events were superimposed on a "walk" to ever increasing diversity. The overall picture then is one far from saturation and far from evolutionary equilibrium. Each new species opens new possibilities for others, i.e., for parasites, hyperparasites, symbionts, predators, etc., and this must lead to an ever increasing trend towards higher diversity. Rosenzweig's (1995) claim that the resource base is still the same and therefore sets limits to species numbers with the implication that, even if newly evolved species provide opportunities for others, the system as a whole cannot "get out of control" by ever-increasing species numbers, is not supported by the overwhelming evidence that many niches are vacant, i.e., that resources are not fully exploited. Also, there is no reason to assume that diversity cannot increase by subdivision of niches. There may well be a limit to diversity determined by the minimum possible body size and population size of species, and by the limited space, energy, minerals etc. available on Earth, but no evidence indicates that we are even close to that limit.

The common occurrence of nonequilibrium conditions in populations and communities has important implications for the significance of interspecific competition. As shown in Chapters 3 and 4, much of the evidence for competition is faulty, although there can be no doubt that it plays an important role in structuring many communities. But, as discussed in Chapter 5, many (and perhaps most) examples given for competition can also be explained, or are indeed better explained, by non-competitive mechanisms. Thus, niche restriction and segregation may simply be a consequence of random selection of niches (habitats, microhabitats, food resources, etc.) in largely empty niche space. The fact that species do not expand into adjacent niche space may be due to the need to specialize. In other words, species need to be adapted to particular niches in order to survive: species that use too many niches may hang on for a while in all of them, but they will be pushed over the rim when environmental conditions become less favourable. Niche restriction may be further enhanced by the necessity to find mating partners, which may be impossible if niches are too wide. Finally, segregation may be the result of reinforcement of reproductive barriers: segregated species cannot hybridise and produce unfit or less fit offspring.

Much effort has gone into ecological modelling; according to some authors, it now appears feasible that supercomputers can be used to develop "a single theoretical model which would be able to describe the entire dynamics of an ecosystem since the first appearance of life in it up till now" (Chowdhury and Stauffer 2004; Stauffer and Chowdhury 2005).

It is important that such models are not only mathematically sound but are based on realistic assumptions. Unfortunately, this is not often the case. Thus, many if not most models are based on equilibrium assumptions. However, models that assume close species packing and strong competition for resources have limited applicability, because, as we have seen, most communities are not saturated with species and individuals. The model developed in the previous section ("What explains the differences between communities?") shows that communities comprising large-bodied species with much vagility and/or species occurring at high densities are closer to equilibrium than small-bodied species with little vagility and/or occurring at low densities. More data from many animal and plant communities are needed to confirm the findings. Of great importance, the question that should be examined is whether keystone species and, more importantly, "keystone functional groups," that is, sets of species that control critical ecosystem processes, are more likely or less likely to be close to equilibrium conditions than other species, and how whole ecosystems are affected by this.

As is implicit in the discussion in Chapter 10, the first principle of the autecological paradigm relates to the nature of species, and the nature of interactions between individuals and the environment, in contrast to the demographic paradigm which relies heavily on equilibrium assumptions. An autecologist will invest much effort in analysing the characteristics and requirements of species, and explain distributional patterns by these characteristics. The autecological approach does not rely on equilibrium assumptions and in fact rejects them, it deserves more attention and may yield important insights if applied to a variety of communities. However, we should examine its general applicability. Is it possible that it applies primarily to slowly evolving taxa and communities with little adaptive potential and little speciation?

The autecological approach is not the only one that may significantly contribute to future developments in ecology. Of foremost importance are the recent attempts by Brown, Gillooly, Allen, and collaborators to explain ecological patterns by the first principles of body size, temperature, and stoichiometry. Temperature was suggested to be a direct cause of latitudinal gradients in species diversity by Rohde (1978a,b, 1992), and for increased viviparity/brooding in benthic marine invertebrates at high latitudes (Thorson's Rule, Rohde 1985). Allen *et al.* (2002) and Gillooly *et al.* (2002) have made remarkable progress in explaining diversity and embryonic development time by such principles (see pp. 11, 161–164). Brown *et al.* (2004) extended this approach to a metabolic theory of ecology. Their approach does not rely on equilibrium assumptions, and can be expected to become increasingly important in the future.

A radically new approach for solving scientific problems in many fields is that pioneered by Stephen Wolfram (Wolfram 2002), i.e., NKS ("New Kind of Science"). Wolfram applied NKS to evolution (see pp. 11–13).

The applicability of NKS to interpreting ecological processes is shown by the following examples, which include the possibility of establishing general ecological "laws," the existence of vacant niches, the significance of interspecific competition, and the causes of latitudinal gradients in species diversity. All of these points are relevant to the discussion of equilibrium and nonequilibrium in ecological systems (see also Rohde 2005a).

Lawton (1999) raised the question of whether general laws are possible in ecology and concluded that there are numerous "laws" in ecology in the sense of widespread, repeatable patterns. However, there are "hardly any laws that are universally true", because patterns depend on the organisms involved and their environment. Even at the level of populations, it is highly unlikely that theory will ever become truly predictive. At the level of communities, there are "painfully few generalisations, let alone rules or laws". Wolfram's Principle of Computational Equivalence provides the theoretical foundation for Lawton's conclusions. As pointed out above, according to the principle, the computations necessary to predict the fate of any complex system require at least as many steps as contained in the system itself: in other words, general predictive laws that permit shortcutting the computational process are impossible in complex ecological systems. This conclusion is not contradicted by the fact that predictions close to predictive "laws" can be made for those characteristics of species and ecological systems that can be explained by first principles of physics, chemistry, and physiology, as recently shown by Gillooly *et al.* (2002), Allen *et al.* (2002), and Brown *et al.* (2004) (see above and pp. 11, 161–164). Such a "metabolic theory of ecology," based on temperature and mass dependence of metabolic rates, has, for example, established quantitative relationships between temperature/mass dependence of developmental rates, mortality rates, and maximum rates of population growth, as well as mass dependence of population density. However, the theory is restricted to relatively simple phenomena, i.e., the effects of allometry, kinetics, and stoichiometry on the biological processing of energy and materials, and even in this domain not all variation is explained (Brown *et al.* 2004). The Principle of Computational Equivalence implies that the theory cannot be developed to address complex relationships between species in communities. The fact that nonequilibrium conditions are much more common than equilibrium ones, as shown in the preceding chapters, makes it even less likely that such laws can ever be found: most nonequilibrial patterns are largely or entirely unpredictable and can therefore not be modelled by simple equations.

Both the approaches of Kauffman (1993) and Wolfram (2002) show that species seldom (if ever) reach global adaptive optima. Since local optima are abundant and perhaps almost infinite in number, an overwhelming majority will remain unoccupied. In other words, there are a vast number of empty niches. It is highly unlikely that species, even if they are closely related, will occupy the same local optimum, because the processes that have led them to this optimum are largely random, i.e., most species will have little chance for interactions, whether positive or negative. Empirical evidence lends strong support to these conclusions (for examples see pp. 72–76). Results using cellular automata put these findings and, with them, the prevalence of nonequilibrium in ecological systems, in a convincing theoretical framework.

As discussed on pp. 152–165, many studies have attempted to find the causes of latitudinal gradients in species diversity. Rohde (1992, 1998a) gave a nonequilibrium explanation, that is, he suggested that the primary cause, supplemented by some secondary ones, is the accelerated speed of evolution at higher temperatures, resulting from greater mutation rates, shorter generation times, and greater speed of selection at higher temperatures in largely empty niche space. More generally, the hypothesis suggests that species diversity in ecosystems is determined by "effective evolutionary time", i.e., the above factors and the time under which systems have existed under more or less constant conditions. The hypothesis has been supported by a considerable number of recent studies (for details see pp. 158–165). In the framework of NKS, the hypothesis makes sense as well: if additions of more and more "programs" (species) in an automaton occur faster at higher temperatures, as indeed shown in several recent studies using DNA (e.g., Wright *et al.* 2003; Martin and McKay 2004), more species will have evolved at higher than at lower temperatures. Even if addition of species is random, some of them may be more complex than those already present, because even quite simple programs may easily lead to complexity (for an example see Figure 11.2). This would explain the observations that some tropical bird species are more colorful and have more intricate song patterns than birds from colder environments, and that many tropical plant species have very conspicuous and colorful flowers. These phenomena do not necessarily result

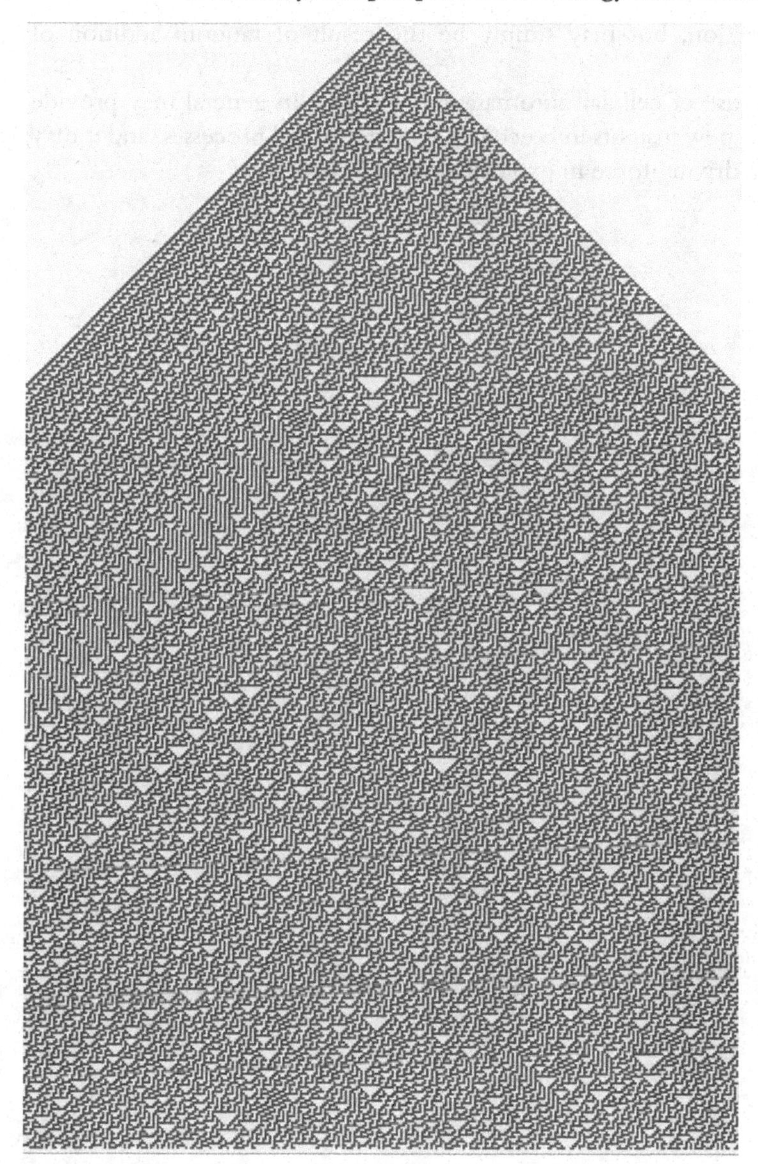

Figure 11.2. An example of a cellular automaton. The evolution of the first few hundred steps of "Rule 30," which states that, if both a cell and its right neighbor were white in the step before, the new colour of the cell should take the previous colour of its left neighbor; if not, the new colour should be its opposite. The left and right lower parts have been truncated. Note some regularities on the left, but randomness (or pseudo-randomness) in most parts of the system. Various statistical and mathematical tests failed to find any regularities in the arrangement of the first million steps below the initial black cell. Based on and modified from Wolfram (2002).

from selection, but may simply be the result of random addition of species.

Future use of cellular automata and of NKS in general may provide significant new insights into ecological patterns and processes, and it may become a driving force in exploring them.

References

Abrams, P. A. (1983). The theory of limiting similarity. *Annual Review of Ecology and Systematics* **14**, 359–376.

Aho, J. M. (1990). Helminth communities of amphibians and reptiles: comparative approaches to understanding patterns and processes. In Esch, G. W., Bush, A. O. and Aho, J. M., eds., *Parasite Communities: Patterns and Processes.* London, Chapman and Hall, pp. 157–195.

Allen, A. P., Brown, J. H. and Gillooly, J. F. (2002). Global biodiversity, biochemical kinetics, and the energetic-equivalence rule. *Science* **297**, 1545–1548.

Anderson, A. N. (1995). Palaeontology. Adaptation and community ecology: a response to Walter and Patterson (1994). *Australian Journal of Ecology* **19**, 241–250.

Andrewartha, H. G. (1970). *Introduction to the Study of Animal Populations*, 2nd edn., London, Chapman and Hall.

Andrewartha, H. G. and Birch, L. C. (1954). *The Distribution and Abundance of Animals.* Chicago, University of Chicago Press.

(1984). *The Ecological Web.* Chicago and London, University of Chicago Press.

Aoyama, J. and Tsukamoto, K. (1997). Evolution of freshwater eels. *Naturwissenschaften* **84**, 17–21.

Aoyama, J., Nishida, M. and Tsukamoto, K. (2001). Molecular phylogeny and evolution in the freshwater eel, genus *Anguilla. Molecular Phylogenetics and Evolution* **20**, 450–459.

Apaloo, J. (2003). Single species evolutionary dynamics. *Evolutionary Ecology* **17**, 33–49.

Arme, C. and Halton, D. W. (1974). Observations on the occurrence of *Diclidophora merlangi* (Trematoda: Monogenea) on the gills of whiting, *Gadus merlangus. Journal of Fish Biology* **4**, 27–32.

Armonies, W. and Reise, K. (2000). Faunal diversity across a sandy shore. *Marine Ecology Progress Series* **196**, 49–57.

Armstrong, R. A. and McGee, R. (1980). Competitive exclusion. *American Naturalist* **115**, 151–170.

Armsworth, P. R. (2002). Recruitment limitation, population regulation and larval connectivity in reef fish metapopulations. *Ecology* cit. Doherty (2002).

Arndt, W. (1940). Der prozentuelle Anteil der Parasiten auf und in Tieren im Rahmen des aus Deutschland bisher bekannten Tierartenbestandes. *Zeitschrift für Parasitenkunde* **121**, 684–689.

Arthur, W. (1982). *The Evolutionary Consequences of Interspecific Competition.* Advances in Ecological Research, London, New York, Academic Press, pp. 127–187.

Bannarescu, P. (1975). *Principles and problems of zoogeography*. Belgrade, NOLIT.

Barker, J. S. F. (1983). Interspecific competition. In Ashburner, M., Carson, H. L. and Thompson, J. N. Jr., eds., *The Genetics and Biology of Drosophila*. London, Academic Press, pp. 285–341.

Barker, R. D. (1987). The diet of herbivores in the sheep rangelands. In Caughley, G., Shepherd, N. and Short, J. eds., *Kangaroos: Their Ecology and Management in the Sheep Rangelands of Australia*. Cambridge, Cambridge University Press, pp. 69–83.

Barraclough, T. G., Harvey, P. H. and Nee, S. (1996). Rate of *rbc*L gene sequence evolution and species diversification in flowering plants (angiosperms). *Proceedings of the Royal Society London B* **263**, 589–591.

Bates, R. M. and Kennedy, C. R. (1990). Interactions between the acanthocephalans *Pomporhynchus laevis* and *Acanthocephalus anguillae* in rainbow trout: testing an exclusion hypothesis. *Parasitology* **100**, 435–444.

Bazin, C., Capy, P., Higuet, D. and Langin, T. (1997). Séquences d'AND mobiles et évolution du génome. Pour Sci., Hors. Sér. Janvier **97**, 106–109. (cit. Harmelin-Vivien 2002).

Begon, M. and Mortimer, M. (1981). *Population Ecology. A Unified Study of Animals and Plants*. Oxford, London, Edinburgh, Boston, Melbourne, Blackwell Scientific.

Begon, M. J., Harper, J. L. and Townsend, C. R. (1990). *Ecology. Individuals, Populations and Communities*, 2nd edn. Boston, Blackwell Scientific.

(1996). *Ecology*. Oxford, Blackwell Scientific.

Ben-Eliahu, M. N. and Safriel, U. N. (1982). A comparison between species diversity of polychaetes from tropical and temperate structurally similar rocky intertidal habitats. *Journal of Biogeography* **9**, 371–390.

Benton, M. J. (1995). Diversification and extinction in the history of life. *Science* **268**, 52–58.

(1998). Analyzing diversification through time: reply to Sepkoski and Miller. *Trends in Ecology and Evolution* **13**, 201.

Benton, M. J. and Pearson, P. N. (2001). Speciation in the fossil record. *Trends in Ecology and Evolution* **16**, 405–411.

Berryman, A. A., (1987). Equilibrium or nonequilibrium: is that the question? *Bulletin of the Ecological Society of America*. **68**, 500–502.

Beveridge, J., Chilton, N. B. and Spratt, D. M. (2002). The occurrence of species flocks in the nematode genus *Cloacina* (Strongyloidea: Cloacininae), parasitic in the stomachs of kangaroos and wallabies. *Australian Journal of Zoology* **50**, 597–620.

Bradley, D. J. (1974). Stability in host-parasite systems. In Usher, M. B. and Williamson, M. H., eds., *Ecological Stability*. London, Chapman and Hall, pp. 71–87.

Bromham, L. and Cardillo, M. (2003). Testing the link between the latitudinal gradient in species richness and rates of molecular evolution. *Journal of Evolutionary Biology* **16**, 200–207.

Bromham, L. D., Rambaut, A. and Harvey, P. H. (1996). Determinants of rate variation in mammalian DNA sequence evolution. *Journal of Molecular Evolution* **43**, 610–621.

Brown, J. H. (1975). Geographical ecology of desert rodents. In Cody, M. L. and Diamond, J. M., eds., *Ecology and Evolution of Communities*. Cambridge, Mass. and London, Belknap Press of Harvard University, pp. 315–341.

(1995). *Macroecology*. Chicago, University of Chicago Press.

(1999). Macroecology: progress and prospect. *Oikos* **87**, 3–14.

Brown, J. H. and Maurer, B. A. (1989). Macroecology: the division of food and space among species on continents. *Science* **243**, 1145–1150.

Brown, J. H., Reichman, O. J. and Davidson, D. W. (1979). Granivory in desert ecosystems. *Annual Review of Ecology and Systematics* **10**, 201–227.

Brown, J. H., Gillooly, J. F., Allen, A. P., Savage, V. M. and West, G. B. (2004). Toward a metabolic theory of ecology. *Ecology* **85**, 1771–1789.

Burr, B. M. and Page, L. M. (1986). Zoogeography of the lower Ohio-upper Mississippi basin. In Hocutt, C. H. and Wiley, E. O., eds., *The Zoogeography of North American Freshwater Fishes*. London, Wiley, pp. 287–324.

Bush, A. O. (1990). Helminth communities in avian hosts: determinants of patterns. In Esch, G. W., Bush, A. O. and Aho, J. M., eds., *Parasite Communities: Patterns and Processes*. London, Chapman and Hall, pp. 197–232.

Butterworth, E. W. and Holmes, J. C. (1984). Character divergence in two species of trematodes (*Pharyngostomoides*: Strigeoidea). *Journal of Parasitology* **70**, 315–316.

Byrnes, T. and Rohde, K. (1992). Geographical distribution and host specificity of ectoparasites of Australian bream, *Acanthopagrus* spp. (Sparidae). *Folia Parasitologica* **39**, 249–264.

Cairns, S. C. and Grigg, G. C. (1993). Population dynamics of red kangaroos (*Macropus rufus*) in relation to rainfall in the South Australian pastoral zone. *Journal of Applied Ecology* **30**, 444–458.

Cannon, L. R. G. (1979). Ecological observations on *Cerithium moniliferum* Kiener (Gastropoda: Cerithiidae) and its trematode parasites at Heron Island, Great Barrier Reef. *Australian Journal of Marine and Freshwater Research* **30**, 365–374.

Cappucino, N. (1995). Novel approaches to the study of population dynamics. In Cappucino, N. and Price, P. W., eds., *Population Dynamics. New Approaches and Synthesis*. San Diego, Academic Press, pp. 3–16.

Cardillo, M. (1999). Latitude and rates of diversification in birds and butterflies. *Proceedings of the Royal Society London* **266**, 1221–1225.

Caswell, H. (1978). Predator-mediated coexistence: a non-equilibrium model. *American Naturalist* **112**, 127–154.

Caswell, H. and Cohen, J. E. (1993). Local and regional regulation of species-area relations: a patch-occupancy model. In Ricklefs, R. E. and Schluter, D., eds., *Species Diversity in Ecological Communities. Historical and Geographical Perspectives*. Chicago, University of Chicago Press, pp. 99–107.

Caughley, G. (1987a). Introduction to the sheep rangelands. In Caughley, G., Shepherd, N. and Short, J., eds., *Kangaroos: Their Ecology and Management in the Sheep Rangelands of Australia*. Cambridge, Cambridge University Press, pp. 1–13.

(1987b). Ecological relationships. In Caughley, G., Shepherd, N. and Short, J., eds., *Kangaroos: Their Ecology and Management in the Sheep Rangelands of Australia*. Cambridge, Cambridge University Press, pp. 159–187.

Caughley, G., Shepherd, N. and Short, J., eds. (1987). *Kangaroos: Their Ecology and Management in the Sheep Rangelands of Australia.* Cambridge, Cambridge University Press.

Chesson, P. L. (1978). Predator-prey theory and variability. *Annual Review of Ecology and Systematics* **9**, 288–325.

(1981). Models for spatially distributed populations: the effect of within-patch variability. *Theoretical Population Biology* **19**, 288–323.

(1982). The stabilizing effect of a random environment. *Journal of Mathematical Ecology* **15**, 1–36.

(1986). Environmental variation and the coexistence of species. In Diamond, J. and Case, T., eds., *Community Ecology*. New York, Harper and Row, pp. 240–256.

Chesson, P. L. and Case, T. J. (1986). Overview: nonequilibrium community theories: chance, variability, history, and coexistence. In Diamond, J. and Case, T., eds., *Community Ecology*. New York, Harper and Row, pp. 229–239.

Chittaro, P. M. and Sale, P. F. (2003). Structure of patch-reef fish assemblages at St. Croix, US Virgin Islands, and One Tree Reef, Australia. *Marine Ecology Progress Series* **249**, 277–287.

Chowdhury, D. and Stauffer, D. (2004). Computer simulations of history of life: speciation, emergence of complex species from simple organisms, and extinctions. *Physica A- Statistical Mechanics and its Applications* **340**, 685–696.

Christiansen, F. B. and Fenchel, T. M. (1977). *Theories of Populations in Biological Communities*. Berlin, Heidelberg, New York, Springer-Verlag.

Clutton-Brock, T. H. and Pemberton, J., eds., (2004). *Soay Sheep. Dynamics and Selection in an Island Population.* Cambridge, Cambridge University Press.

Clutton-Brock, T. H., Price, O. F., Albon, S. D. and Jewell, F. A. (1991). Persistent instability and population regulation in Soay sheep. *Journal of Animal Ecology* **60**, 593–608.

Cody, M. L. (1974). *Competition and the Structure of Bird Communities*. Princeton, N. J., Princeton University Press.

Cody, M. L. and Diamond, J. M., eds., (1975). *Ecology and Evolution of Communities*. Cambridge, Mass. and London, Belknap Press of Harvard University.

Colwell, R. K. (1984). What's new? Community ecology discovers biology. In Price, P. W., Slobodchikoff, C. N. and Gaud, W. S., eds., (1984). *A New Ecology. Novel Approaches to Interactive Systems*. New York, John Wiley & Sons, pp. 387–396.

Coman, B. J. (1996). A short history of the rabbit in Australia. *Quadrant* **4**, 21–26.

Combes, C. (2001). *Parasitism. The Ecology and Evolution of Intimate Interactions*. Chicago and London, University of Chicago Press.

Compton, S. G., Lawton, J. H. and Rashbrook, V. K. (1989). Regional diversity, local community structure and vacant niches: the herbivorous arthropods of bracken in South Africa. *Ecological Entomology* **14**, 365–373.

Connell, J. H. (1975). Some mechanisms producing structure in natural communities: a model and evidence from field experiments. In Cody, M. L. and Diamond, J. M., eds., *Ecology and Evolution of Communities*. Cambridge, Mass., Harvard University Press, pp. 460–490.

(1978). Diversity in tropical rain forests and coral reefs. *Science* **199**, 1302–1309.

(1979). Tropical rain forests and coral reefs as open non-equilibrium systems. *Symposia of the British Ecological Society* **20**, 141–163.

(1980). Diversity and the coevolution of competitors, or the ghost of competition past. *Oikos* **35**, 131–138.

(1983). On the prevalence and relative importance of interspecific competition: evidence from field experiments. *American Naturalist* **122**, 661–696.

Connell, J. H. and Green, P. T. (2000). Seedling dynamics over thirty-two years in a tropical rain forest tree. *Ecology* **81**, 568–584.

Connell, J. H., Tracey, J. G. and Webb, L. J. (1984). Compensatory recruitment, growth, and mortality as factors maintaining rain forest tree diversity. *Ecological Monographs* **54**, 141–184.

Connor, E. F. and Simberloff, D. S. (1978). Species number and compositional similarity of the Galapagos flora and avifauna. *Ecological Monographs* **48**, 219–248.

(1986). Competition, scientific method, and null models in ecology. *American Scientist* **74**, 155–162.

Cooper, G. (1993). The competition controversy in community ecology. *Biology and Philosophy* **8**, 359–384.

(2001). Must there be a balance in nature? *Biology and Philosophy* **16**, 481–506.

Cornell, H. V. (1985a). Local and regional richness of cynipine gall wasps on California oaks. *Ecology* **66**, 1247–1260.

(1985b). Species assemblages of cynipid gall wasps are not saturated. *American Naturalist* **126**, 565–569.

(1999). Unsaturation and regional influences on species richness in ecological communities: a review of the evidence. *Ecoscience* **6**, 303–315.

(2001). Diversity, community/regional level. In Levin, S., ed., *Encyclopedia of Biodiversity*, vol. 32. New York, Academic Press, pp. 161–177.

Cornell, H. V. and Karlson, R. H. (1997). Local and regional processes as controls of species richness. In Tilman, D. and Kareiva, P., eds., *Spatial Ecology. The Role of Space in Population Dynamics and Interspecific Interactions*. Princeton, N. J., Princeton University Press, pp. 250–268.

Cornell, H. V. and Lawton, J. H. (1992). Species interactions, local and regional processes, and limits to the richness of ecological communities: a theoretical perspective. *Journal of Animal Ecology* **61**, 1–12.

Courtillot, V. and Gaudemer, Y. (1996). Effects of mass extinctions on biodiversity. *Nature* **381**, 146–148.

Crame, J. A. (2001). Taxonomic diversity gradients through geological time. *Diversity and Distributions* **7**, 175–189.

Crawley, M. J., ed., (1986). *Plant Ecology*. Oxford, Blackwell Scientific Publishing.

Crawley, M. J. (1990). The population dynamics of plants. In Hassell, M. P. and May, R. M., eds., *Population, Regulation and Dynamics. Philosophical Transactions of the Royal Society of London, Series B* **330**, 125–140.

Criddle, R. S., Church, J. N., Smith, B. N. and Hansen, L. D. (2003). Fundamental causes of the global patterns of species range and richness. *Russian Journal of Plant Physiology* **50**, 192–199.

Currie, D. J. (1991). Energy and large-scale patterns of animal and plant-species richness. *American Naturalist* **137**, 27–49.

Currie, D. J. and Paquin, V. (1987). Large scale biogeographical patterns of species richness of trees. *Nature* **329**, 326–327.

Curtis, L. A. and Hubbard, K. M. K. (1993). Species relationships in a marine gastropod-trematode ecological system. *Biological Bulletin* **184**, 25–35.

Darwin, C. (1859). *On the Origin of Species by Means of Natural Selection*. Facsimile 1964, Cambridge, Mass., Harvard University Press.

Dayton, P. K. (1971). Competition, disturbance, and community organization: The provision and subsequent utilization of space in a rocky intertidal community. *Ecological Monographs* **41**, 351–389.

DeAngelis, D. L. and Waterhouse, J. C. (1987). Equilibrium and nonequilibrium concepts in ecological models. *Ecological Monographs* **57**, 1–21.

Den Boer, P. J. and Reddingius, J. (1996). *Regulation and Stabilization Paradigms in Population Ecology*. London, Chapman and Hall.

Dennis, B. and Taper, B. (1994). Density dependence in time series observations of natural populations: estimation and testing. *Ecological Monographs* **64**, 205–224.

Diamond, J. (1973). Distributional ecology of New Guinea birds. *Science* **179**, 759–769.

(1975). Assembly of species communities. In Cody, M. and Diamond, J., eds., *Ecology and Evolution of Communities*. Cambridge, Mass., Harvard University Press, pp. 342–344.

Diamond, J. and Case, T. J., eds., (1986). *Community Ecology*. New York, Harper and Row.

Dickman, C. R., Pressey, R. L., Lim, L. and Parnaby, H. E. (1993). Mammals of particular conservation concern in the Western Division of New South Wales. *Biological Conservation* **65**, 219–248.

Dobshansky, T. (1957). Discussion, from Andrewartha, H. G.: the use of conceptual models in population ecology. *Cold Spring Harbour Symposia on Quantitative Biology*, p. 235.

Dobson, S. I., Arnott, S. E. and Cottingham, K. L. (2000). The relationship in lake communities between primary productivity and species richness. *Ecology* **81**, 2662–2679.

Doherty P. J. (2002). Variable replenishment and the dynamics of reef fish populations. In Sale, P. F., ed., *Coral Reef Fishes. Dynamics and Diversity in a Complex Ecosystem*. Amsterdam, Academic Press, pp. 327–355.

Doherty, P. J. and Williams, D. M. (1988). The replenishment of coral reef fish populations. *Oceanography and Marine Biology* **26**, 487–551.

Dunham, A. E. (1980). An experimental study of interspecific competition between the iguanid lizards *Sceloporus merriami* and *Urosaurus ornatus*. *Ecological Monographs* **50**, 309–330.

Eadie, J. M. (1987). Size ratios and artifacts: Hutchinson's rule revisited. *American Naturalist* **129**, 1–17.

Edmunds, J., Cushing, J. M., Constantino, R. F., Henson, S. M., Dennis, B. and Desharnais, R. A. (2003). Park's *Tribolium* competition experiments: a non-equilibrium species coexistence hypothesis. *Journal of Animal Ecology* **72**, 703–712.

Egerton, F. N. (1973). Changing concepts of the balance of nature. *The Quarterly Review of Biology* **48**, 322–350.

Ehrlich, P. R., Ehrlich, A. H. and Holdren, J. P. (1977). *Ecoscience. Population, Resources, Environment*. San Francisco, W. H. Freeman.

Elder, H. Y. (1979). Studies on the host-parasite relationship of the parasitic prosobranch *Thyca cristallina* and the asteroid starfish *Linckia laevigata*. *Journal of Zoology* **187**, 369–391.

Elton, C. (1933). *The Ecology of Animals*. New York, MacMillan.

Elton, C. S. (1946). Competition and the structure of ecological communities. *Journal of Animal Ecology* **15**, 54–68.

Emison, W. B., Beardsell, C. M. and Temby, I. D. (1994). The biology and status of the long-billed Corella in Australia. *Proceedings of the Western Foundation of Vertebrate Zoology* **5**, 211–247.

Esch, G. E. and Fernandez, J. C. (1993). *A Functional Biology of Parasitism*. London, Chapman and Hall.

Esch, G. E., Bush, A. and Aho, J. (1990). *Parasite Communities: Patterns and Processes*. London, New York, Chapman and Hall.

Farrell, E. J. (1998). *Fasciola hepatica* in the New England Region of New South Wales. BSc. Honours thesis, University of New England, Armidale, Australia.

Fauth, J. E., Bernardo, J., Camara, M., Resetarits, J. Jr., van Buskirk, J. and McCollum, S. A. (1996). Simplifying the jargon of community ecology: a conceptual approach. *American Naturalist* **147**, 282–286.

Fenchel, T., ed., (1999). *Ecology 1999 and Tomorrow*. Copenhagen, Nordic Ecological Society Oikos.

Flessa, K. W. (1975). Area, continental drift and mammalian diversity. *Paleobiology* **1**, 189–194.

Forrester, C. E., Vance, R. R. and Steele, M. A. (2002). Simulating large-scale population dynamics using small-scale data. In Sale, P. F., ed., (2002). *Coral Reef Fishes. Dynamics and Diversity in a Complex Ecosystem*. Amsterdam, Academic Press, pp. 275–301.

Francis, A. P. and Currie, D. J. (1998). Global patterns of tree species richness in moist forests: another look. *Oikos* **81**, 598–602.

(2003). A globally consistent richness-climate relationship for angiosperms. *American Naturalist* **161**, 523–536.

Freeman, S. and Herron, J. C. (2004). *Evolutionary Analysis*. 3rd edn. Upper Saddle River, NJ, Pearson Prentice Hall.

Fried, B. and Diaz, V. (1987). Site-finding and pairing of *Echinostoma revolutum* (Trematoda) on the chick chorioallantois. *Journal of Parasitology* **73**, 546–548.

Fried, B., Tancer, R. B. and Fleming, S. J. (1980). *In vitro* pairing of *Echinostoma revolutum* (Trematoda) metacercariae and adults, and characterisation of worm products involved in chemoattraction. *Journal of Parasitology* **66**, 1014–1018.

Gaston, K. J. and Blackburn, T. M. (1999). A critique of macroecology. *Oikos* **84**, 353–368.

Gause, G. F. (1935). *The Struggle for Existence*. Baltimore, Williams & Wilkins.

Geets, A., Coene, H. and Ollevier, F. (1997). Ectoparasites of the whitespotted rabbitfish, *Siganus sutor* (Valenciennes, 1835) off the Kenyan coast: distribution within the host population and site selection on the gills. *Parasitology* **115**, 69–79.

Geller, W. and Gude, H. (1989). Lake Constance – the largest German lake. In Lampert, W. and Rothaupt, K. O., eds., *Limnology in the FRG*. 24th Congress of the International Association of Theoretical and Applied Limnology.

Gilbert, F. S. (1980). The equilibrium theory of island geography: fact or fiction. *Journal of Biogeography* **7**, 209–235.

Giller, P. S. and Gee, J. H. R. (1987). The analysis of community organization: the influence of equilibrium, scale and terminology. In Gee, J. H. R. and Giller, P. S., eds., *Organization of Communities Past and Present*. Oxford, Blackwell Scientific, pp. 519–542.

Gillooly, J. F., Charnov, E. L., West, G. B., Savage, V. M. and Brown, J. H. (2002). Effects of size and temperature on developmental time. *Nature* **417**, 70–73.

Gleason, H. A. (1926). The individualistic concept of the plant association. *Bulletin of the Torrey Botanical Club* **53**, 7–26.

Goater, T. M., Esch, G. W. and Bush, A. O. (1987). Helminth parasites of sympatric salamanders: ecological concepts at infracommunity, component and compound community levels. *American Midland Naturalist* **118**, 289–300.

Godfray, H. C. J. (1994). *Parasitoids: Behavioural and Evolutionary Ecology*. Princeton, N. J., Princeton University Press.

Godfray, H. C. J. and Lawton, J. H. (2001). Scale and species numbers. *Trends in Ecology and Evolution* **16**, 400–404.

Gosper D. M. (1992) The ecology of parasites of *Anguilla reinhardtii* (long-finned eel) and *Anguilla australis* (short-finned eel). BSc Honours thesis, University of New England, Armidale, Australia.

Gotelli, N. J. and Graven, G. R. (1996). *Null Models in Ecology*. Washington, D.C., Smithsonian Institution Press.

Gotelli, N. J. and McCabe, D. J. (2002). Species co-occurrence: a meta-analysis of J. M. Diamond's assembly rules model. *Ecology* **83**, 2091–2096.

Gotelli, N. J., and Rohde, K. (2002). Co-occurrence of ectoparasites of marine fishes: null model analysis. *Ecology Letters* **5**, 86–94.

Grant, P. R. (1972). Convergent and divergent character displacement. *Biological Journal of the Linnaean Society* **4**, 39–68.

(1975). The classical case of character displacement. *Evolution Biologica* **8**, 237–337.

Grant, P. R. and Schluter, D. (1984). Interspecific competition inferred from patterns of guild structure. In Strong, D. R. Jr., Simberloff, D., Abele, L. G. and Thistle, A. B., eds., *Ecological Communities: Conceptual Issues and the Evidence*. Princeton, N. J., Princeton University Press, pp. 201–231.

Grime, J. P. (1979). *Plant Strategies and Vegetation Processes*. London, Wiley.

Grinnell, J. (1904). The origin and distribution of the chestnut-backed chickadee. *Auk* **21**, 364–382.

Hanski, I. (1997). Predictive and practical metapopulation models: the incidence function approach. In Tilman, D. and Kareiva, P. (1997). *Spatial Ecology: the Role of Space in Population Dynamics and Interspecific Interactions*. Princeton N. J., Monographs in Population Biology, no. 30. Princeton University Press, pp. 21–45.

(1999). *Metapopulation Ecology*. Oxford, Oxford University Press.

Hanski, I. and Gilpin, M. E. (1991). Metapopulation dynamics: a brief history and conceptual domain. In Gilpin, M. and Hanski, I., eds., *Metapopulation Dynamics: Empirical and Theoretical Investigations*. London, Academic Press, pp. 3–16.

Hanski, I. and Kuussaari, M. (1995). Butterfly metapopulation dynamics. In Cappucino, N. and Price, P. W., eds., *Population Dynamics. New Approaches and Synthesis*. San Diego, Academic Press, pp. 149–171.

Hanski, I. and Ovaskainen, O. (2002). Extinction debt at extinction threshold. *Conservation Biology* **16**, 666–673.

Hanski, I. I., Woiwod, I. and Perry, J. (1993). Density dependence, population persistence, and largely futile arguments. *Oecologia* **95**, 595–598.

Harmelin-Vivien, M. L. (2002). Energetics and fish diversity on coral reefs. In Sale, P. F., ed., *Coral Reef Fishes. Dynamics and Diversity in a Complex Ecosystem*. Amsterdam, Academic Press, pp. 265–274.

Harris, G. P. (1986). *Phytoplankton Ecology. Structure, Function and Fluctuation*. London, Chapman and Hall.

Hassell, M. P. and May, R. M., eds., (1990). *Population, Regulation and Dynamics*. Philosophical Transactions of the Royal Society of London, Series B **330**, 121–304.

Haukisalmi, V. and Henttonen, H. (1993a). Coexistence in helminths of the bank vole *Clethrionomys glareolus*. I. Patterns of co-occurrence. *Journal of Animal Ecology* **62**, 221–229.

(1993b). Coexistence in helminths of the bank vole *Clethrionomys glareolus*. II. Intestinal distribution and interspecific interactions. *Journal of Animal Ecology* **62**, 230–238.

Hawkins, A. (1993). Complex interactions between dispersal and dynamics: lessons from coupled logistic equations. *Ecology* **74**, 1362–1372.

Hawkins, B. A. and Compton, S. G. (1992). African fig wasp communities: undersaturation and latitudinal gradients in species richness. *Journal of Animal Ecology* **61**, 361–372.

Hawkins, B. A., Field, R., Cornell, H. V. *et al.* (2003). Energy, water, and broad-scale geographic patterns of species richness. *Ecology* **84**, 3105–3117.

Hayward, C. J., Perera, K. M. L. and Rohde, K. (1998). Assemblages of ectoparasites of a pelagic fish, slimy mackerel (*Scomber australasicus*) from Southeastern Australia. *International Journal for Parasitology* **28**, 263–273.

Hendrickson, H. T. (1978). Sympatric speciation: evidence? *Science* **200**, 345–346.

Hengeveld, R. (1994). Biodiversity – the diversification of life in a non-equilibrium world. *Biodiversity Letters* **2**, 1–10.

Hengeveld, R. and Walter, G. H. (1999). The two coexisting ecological paradigms. *Acta Biotheoretica* **47**, 141–170.

Hillebrand, H. (2004). On the generality of the latitudinal diversity gradient. *American Naturalist* **163**, 192–211.

Hine P. M. (1978) Distribution of some parasites of freshwater eels in New Zealand. *New Zealand Journal of Marine and Freshwater Research* **12**, 179–187.

(1980a) Distribution of helminths in the digestive tracts of New Zealand freshwater eels. 1. Distribution of digeneans. *New Zealand Journal of Marine and Freshwater Research* **14**, 329–338.

(1980b) Distribution of helminths in the digestive tracts of New Zealand freshwater eels. 2. Distribution of nematodes. *New Zealand Journal of Marine and Freshwater Research* **14**, 339–347.

Hine P. M., Francis R. I. C. C. (1980) Distribution of helminths in the digestive tracts of New Zealand freshwater eels. 3. Interspecific associations and conclusions. *New Zealand Journal of Marine and Freshwater Research* **14**, 349–356.

Hixon, M. A. and Webster, M. S. (2002). Density dependence in reef fish populations. In Sale, P. F., ed., *Coral Reef Fishes. Dynamics and Diversity in a Complex Ecosystem*. Amsterdam, Academic Press, pp. 303–325.

Holbrook, S. J. and Schmitt, R. J. (2002). Competition for shelter space causes density-dependent predation mortality in damselfishes. *Ecology* **83**, 2855–2868.

Holmes, J. C. (1973). Site selection by parasitic helminths: interspecific interactions, site segregation, and their importance to the development of helminth communities. *Canadian Journal of Zoology* **51**, 333–347.

(1987). The structure of helminth communities. *International Journal for Parasitology* **17**, 203–208.

(1990). Helminth communities in marine fishes. In Esch, G., Bush, A. and Aho, J., eds., *Parasite Communities: Patterns and Processes*. London, New York, Chapman and Hall, pp. 101–130.

Holmes, J. C. and Price, P. W. (1986). Communities of parasites. In Kikkawa, J. and Anderson, D. J., eds., *Community Ecology: Pattern and Process*. Melbourne, Blackwell Scientific Publishing, pp. 187–213.

Holyoak, M. and Lawton, J. H. (1993). Comment arising from a paper by Wolda and Dennis: using and interpreting the results of tests for density dependence. *Oecologia* **95**, 592–594.

Hubbell, S. P. (1980). Seed predation and the coexistence of tree species in tropical forests. *Oikos* **35**, 214–229.

(2001). *The Unified Neutral Theory of Biodiversity and Biogeography*. Princeton, N. J., Princeton University Press.

Hubbell, S. P. and Foster, R. B. (1986). Biology, chance, and history and the structure of tropical rain forest tree communities. In Diamond, J. and Case, T. J., eds., *Community Ecology*. New York, Harper and Row, pp. 314–329.

Hubbell, S. P., Condit, R. and Foster, R. B. (1990). Presence and absence of density dependence in a neotropical tree community. *Philosophical Transactions of the Royal Society London Series B* **330**, 269–281.

Hubbell, S. P., Foster, R. B., O'Brien, S. T., Harms, K. E., Condit, R., Weschler, B., Wright, S. J. and Loo de Lao, S. (1999). Light gap disturbances, recruitment limitation, and tree diversity in a neotropical forest. *Science* **283**, 554–557.

Huffaker, C. B. (1958). Experimental studies on predation: dispersion factors and predator–prey oscillations. *Hilgardia* **27**, 343–383.

Hughes, L. (2003). Climate change and Australia: trends, projections and impacts. *Australian Ecology* **28**, 423–443.

Hughes, T. P., Bellwood, D. R. and Connolly, S. R. (2002). Biodiversity hotspots, centres of endemicity, and the conservation of coral reefs. *Ecology Letters* **5**, 775–784.

Hutchinson, G. E. (1948). Circular causal systems in ecology. *Annals of the New York Academy of Sciences* **50**, 221–246.

(1957). Concluding remarks. *Cold Spring Harbour Symposium on Quantitative Biology* **22**, 415–427.

(1959). Homage to Santa Rosalia, or why are there so many kinds of animals? *American Naturalist* **93**, 145–159.

(1961). The paradox of the plankton. *American Naturalist* **95**, 137–145.

Jablonski, D. (1991). Extinctions: a paleontological perspective. *Science* **253**, 754–757.

(1993). The tropics as a source of evolutionary novelty through geological time. *Nature* **364**, 142–144.

(1999). The future of the fossil record. *Science* **284**, 2114–2116.

Jablonski, D. and Sepkoski, J. J. Jr. (1996). Paleobiology, community ecology, and scales of ecological pattern. *Ecology* **77**, 1367–1378.

Jablonski, D., Roy, K., Valentine, J. W., Price, R. M. and Anderson, P. S. (2003). The impact of the pull of the Recent on the history of marine diversity. *Science* **300**, 1133–1135.

Jackson, J. B. C. and Johnson, K. G. (2001). Measuring past biodiversity. *Science* **293**, 2401–2404.

Janovy, J., Jr., Clopton R. E. and Percival, T. J. (1992). The roles of ecological and evolutionary influences in providing structure to parasite species assemblages. *Journal of Parasitology* **78**, 630–640.

Janovy, J., Jr., Clopton R. E., Clopton, D. A., Snyder, S. D., Efting, A. and Krebs, L. (1995). Species density distributions as null models for ecologically significant interactions of parasite species in an assemblage. *Ecological Modelling* **77**, 189–196.

Jarkovsky, J., Morand, S. and Simková, A. (2004). Reproductive barriers between congeneric monogenean parasites (*Dactylogyrus*: Monogenea): attachment apparatus morphology or copulatory organ incompatibility? *Parasitology Research* **92**, 95–105.

Källander, H. (1981). The effects of provision of food in winter on a population of the Great Tit *Parus major* and the Blue Tit *P. caeruleus. Ornis Scandinaviae* **12**, 244–248.

Kamegai, S. (1986). Studies on *Diplozoon nipponicum* Goto, 1891. The gathering phenomenon of diporpae and the effect of cortisone acetate on the union of diporpae. In Howell, M. J., ed., *Parasitology-Quo Vadit 2*. Handbook, program and abstracts. 6th International Congress of Parasitology, Australian Academy of Science, Canberra, p. 161.

Kareiva, P. (1994). Diversity begets productivity. *Nature* **368**, 686–687.

Kaspari, M., Ward, P. S. and Yuan, M. (2004). Energy gradients and the geographical distribution of local ant diversity. *Oecologia* **140**, 407–413.

Kauffman, S. A. (1993). *The Origins of Order. Self-organization and Selection in Evolution*. New York, Oxford, Oxford University Press.

Kawano, K. (2002). Character displacement in giant rhinoceros beetles. *American Naturalist* **159**, 255–271.

Kearn, G. C. (1970). The production, transfer and assimilation of spermatophores by *Entobdella soleae*, a monogenean kin parasite of the common sole. *Parasitology* **60**, 301–311.

(1988). The monogenean skin parasite *Entobdella soleae*: movement of adults and juveniles from host to host (*Solea solea*). *International Journal for Parasitology* **18**, 313–319.

Kennedy, C. R. (1985). Site segregation by species of Acanthocephala in fish, with special reference to eels, *Anguilla anguilla. Parasitology* **90**, 375–390.
(1990). Helminth communities in freshwater fish: structured communities or stochastic assemblages. In Esch, G., Bush, A. O. and Aho, J. M., eds., *Parasite Communities: Patterns and Processes.* London, New York, Chapman and Hall, pp. 131–156.
(1992). Field evidence for interactions between the acanthocephalans *Acanthocephalus anguillus* and *A. luci* in eels *Anguilla anguilla. Ecological Parasitology* **11**, 122–134.
(1995). Richness and diversity of macroparasite communities in tropical eels *Anguilla reinhardtii* in Queensland, Australia. *Parasitology* **111**, 233–245.
Kennedy, C. R. and Guégan, J. F., (1994). Regional versus local helminth parasite richness in British freshwater fish: saturated or unsaturated parasite communities? *Parasitology* **109**, 175–185.
(1996). The number of niches in intestinal helminth communities of *Anguilla anguilla*: are there enough spaces for parasites? *Parasitology* **113**, 293–302.
Kennedy, C. R., Di Cave, D., Berrilli, F. and Orecchia, P. (1997). Composition and structure of helminth communities in eels *Anguilla anguilla* from Italian lagoons. *Journal of Helminthology* **71**, 35–40.
Keymer, A. E. (1982). Density-dependent mechanisms in the regulation of intestinal helminth populations. *Parasitology* **84**, 573–587.
Kiener, A. and Richard-Vindard, G. (1972). Fishes of the continental waters of Madagascar. In Battistini, R. and Richard-Vindard, G., eds., *Biogeography and Ecology in Madagascar.* The Hague, W. Junk B.V. Publishing, pp. 477–499.
Kimura, M. (1983). *The Neutral Theory of Molecular Evolution.* New York, Cambridge University Press.
Kingsland, S. (1995). *Modeling Nature*, 2nd edn. Chicago, University of Chicago Press.
Koch, M. (2003). Faunal survey. II. The distribution of digenean trematodes within the New England Tablelands. Memoirs of the Queensland Museum.
Kormondy, E. J. (1969). *Concepts of Ecology*. Englewood Cliffs, N. J., Prentice-Hall.
Körner, C. (2000). Why are there global gradients in species richness? Mountains might hold the answer. *Trends in Ecology and Evolution* **15**, 513–514.
Koszowski, J. (1999). Adaptation: a life history perspective. *Oikos* **86**, 185–194.
Krebs, C. J. (1985). *Ecology. The Experimental Analysis of Distribution and Abundance.* New York, Harper and Row.
(1991). The experimental paradigm and long-term population studies. *Ibis* **133**, 3–8.
(2001). *Ecology: the Experimental Analysis of Distribution and Abundance.* 5th edn. San Francisco, Addison Wesley.
Kube, J., Kube, S. and Dierschke, V. (2002). Spatial and temporal variations in the trematode component community of the mudsnail *Hydrobia ventrosa* in relation to the occurrence of waterfowl as definitive hosts. *Journal of Parasitology* **88**, 1075–1086.
Kuris, A. M. (1990). Guild structure of larval trematodes in molluscan hosts: prevalence, dominance and significance of competition. In Esch, G., Bush, A. and Aho, J., eds., *Parasite Communities: Patterns and Processes.* London, Chapman and Hall, pp. 69–100.

Kuris, A. M. and Lafferty, K. D. (1994). Community structure: larval trematodes in snail hosts. *Annual Review of Ecology and Systematics* **25**, 189–217.

Lafferty, K. D., Sammond, D. T. and Kuris, A. M. (1994). Analysis of larval trematode communities. *Ecology* **75**, 2275–2285.

Lambert, A. and Maillard, C. (1974). Parasitisme branchial simultane par deux especies de *Diplectanum* Diesing, 1858 (Monogenea, Monopisthocotylea) chez *Dicentrarchus labrax* (L., 1758) (Teleosteen). *Comptes Rendus Academie de Sciences, Paris*, **279** (D), 1345–1347.

(1975). Repartition branchiale de deux monogenes: *Diplectanum aequans* (Wagener 1857) Diesing, 1858 et *Diplectanum laubieri* Lambert et Maillard, 1974 (Monogenea, Monopisthocotylea) parasites simultanes de *Dicentrarchus labrax* (teleosteen). *Annales de Parasitologie Humaine et Comparee* **50**, 691–699.

Latham, R. E. and Ricklefs, R. E. (1993). Global patterns of tree species richness in moist forests: energy-diversity theory does not account for variation in tree species richness. *Oikos* **67**, 325–333.

Lawton, J. H. (1982). Vacant niches and unsaturated communities: a comparison of bracken herbivores at sites on two continents. *Journal of Animal Ecology* **51**, 573–595.

(1984a). Herbivore community organization: general models and specific tests with phytophagous insects. In Price, P. W., Slobodchikoff, C. N. and Gaud, W. S., eds., *A New Ecology. Novel Approaches to Interactive Systems*. New York, John Wiley & Sons, pp. 329–352.

(1984b). Non-competitive populations, non-convergent communities, and vacant niches: the herbivores of bracken. In Strong, D. R. Jr., Simberloff, D., Abele, L. G. and Thistle, A. B., eds., *Ecological Communities: Conceptual Issues and the Evidence*. Princeton, N. J., Princeton University Press, pp. 67–101.

(1999). Are there general laws in ecology? *Oikos* **84**, 177–192.

(2000). *Community Ecology in a Changing World*. Nordbünte, Oldendorf, Ecology Institute.

Lawton, J. H., and MacGarvin, M. (1986). The organization of herbivore communities. In Kikkawa, J. and Anderson, D. J., eds., *Community Ecology: Pattern and Process*. Melbourne, Blackwell Scientific Publishing, pp. 163–186.

Lawton, J. H. and Strong, D. R. (1981). Community patterns and competition in folivorous insects. *American Naturalist* **118**, 317–338.

Lawton, J. H., Lewinsohn, T. M. and Compton, S. G. (1993). Patterns of diversity for insect herbivores on bracken. In Ricklefs, R. E. and Schluter, D., eds., *Species Diversity in Ecological Communities. Historical and Geographical Perspectives*. Chicago, University of Chicago Press, pp. 178–184.

Lello, J., Boag, B., Fenton, A., Stevenson, I. R. and Hudson, P. J. (2004). Competition and mutualism among the gut helminths of a mammalian host. *Nature* **428**, 840–845.

Leslie, P. H., Park, T. and Mertz, D. B. (1968). The effect of varying the initial numbers on the outcome of competition between two *Tribolium* species. *Journal of Animal Ecology* **37**, 9–23.

Lester, R. J. G. and Adams, J. R. (1974). *Gyrodactylus alexanderi*: reproduction, mortality, and effect on its host *Gasterosteus aculeatus*. *Canadian Journal of Zoology* **52**, 827– 833.

Levin, S. A. (1970). Community equilibria and stability, and an extension of the competitive exclusion principle. *American Naturalist* **104**, 413–423.

(1992). The problem of pattern and scale in ecology. *Ecology* **73**, 1943–1967.

(1998). Ecosystems and the biosphere as complex adaptive systems. *Ecosystems* **1**, 431–436.

(2000). Multiple scales and the maintenance of biodiversity. *Ecosystems* **3**, 498–506.

Levin, S. A. and Paine, R. T. (1974). Disturbance, patch formation, and community structure. *Proceedings of the National Academy of Science USA* **71**, 2744–2747.

Levin, S. A. and Pimentel, D. (1981). Selection of intermediate rates of increase in parasite–host systems. *American Naturalist* **117**, 308–315.

Likens, G. E. (1975). Primary production of inland ecosystems. In Lieth, H. and Whittaker, R. H., eds., *Primary Productivity in the Biosphere.* Berlin, Springer Verlag, pp. 185–202.

Lindström, J., Lundberg, P., Ranta, E. and Kaitala, V. (1999). Oikos, 50 years of ecology. *Oikos* **87**, 462–475.

Littlewood, D. T. J., Rohde, K., Bray, R. A. and Herniou, E. A. (1999). Phylogeny of the Platyhelminthes and the evolution of parasitism. *Biological Journal of the Linnaean Society* **68**, 257–287.

Loveridge, A. J. and Macdonald, D. W. (2003). Niche separation in sympatric jackals (*Canis mesomelas* and *Canis adustus*). *Journal of Zoology (London)* **259**, 143–153.

Lowman, M. D. (1985). Temporal and spatial variability in insect grazing of the canopies of five Australian rainforest species. *Australian Journal of Ecology* **10**, 7–24.

Lozano, S. and Zapata, F. A. (2003). Short-term temporal patterns of early recruitment of coral reef fishes in the tropical eastern Pacific. *Marine Biology* **142**, 399–409.

MacArthur, R. H. (1972). *Geographic Ecology: Patterns in the Distribution of Species.* New York, Harper and Row.

MacArthur, R. H. and Levins, R. (1967). The limiting similarity, convergence and divergence of coexisting species. *American Naturalist* **101**, 377–385.

MacArthur, R. H. and Wilson, E. O. (1963). An equilibrium theory of insular zoogeography. *Evolution* **17**, 373–387.

(1967). *The Theory of Island Biogeography*. Princeton, N. J., Princeton University Press.

Mahon, R. (1984). Divergent structure in fish taxocenes of north temperate streams. *Canadian Journal of Fisheries and Aquatic Science* **41**, 330–350.

Marcogliese, D. J. and Cone, D. K. (1993) What metazoan parasites tell us about the evolution of American and European eels. *Evolution* **47**, 1632–1635.

(1996) On the distribution and abundance of eel parasites in Nova Scotia: influence of pH. *Journal of Parasitology* **82**, 389–399.

(1998). Comparison of richness and diversity of macroparasite communities among eels from Nova Scotia, the United Kingdom and Australia. *Parasitology* **116**, 73–83.

Martin, P. R. and McKay, J. K. (2004). Latitudinal variation in genetic divergence of populations and the potential for future speciation. *Evolution* **58**, 938–945.

Martin, A. P and Palumbi, S. R. (1993). Body size, metabolic rate, generation time, and the molecular clock. *Proceedings of the National Academy of Science USA* **90**, 4087–4091.

Martins, R. P., Lewinsohn, T. M. and Lawton, J. H. (1995). First survey of insects feeding on *Pteridium aquilinum* in Brazil. *Rev. Bras. Entomol.* **39**, 151–156.

May, R. M. (1973). *Stability and Complexity in Model Ecosystems.* Princeton, N. J., Princeton University Press.

(1975). Deterministic models with chaotic dynamics. *Nature* **261**, 165–166.

(1981). The role of theory in ecology. *American Zoologist* **21**, 903–910.

(1986). The search for patterns in the balance of nature: advances and retreats. *Ecology* **67**, 1115–1126.

May, R. M. and Anderson, R. M. (1978). Regulation and stability of host-parasite population interactions. II. Destabilizing processes. *Journal of Animal Ecology* **47**, 249–267.

May, R. M. and MacArthur, R. H. (1972). Niche overlap as a function of environmental variability. *Proceedings of the National Academy of Sciences USA* **69**, 1109–1113.

Maynard Smith, J. (1974). *Models in Ecology.* Cambridge, Cambridge University Press.

Mayr, E. (1976). Bird speciation in the tropics. In Mayr, E., ed., *Evolution and the Diversity of Life.* Cambridge Mass. and London, Belknap Press, pp. 176–187.

McGill, B. J. (2003). A test of the unified neutral theory of biodiversity. *Nature* **422**, 881–885.

McGlone, M. S. (1996). When history matters: scale, time, climate and tree diversity. *Global Ecology and Biogeography Letters* **5**, 309–314.

McIntosh, R. P. (1987). Pluralism in ecology. *Annual Review of Ecology and Systematics* **18**, 321–341.

Middleton, D. (1993). Persistence of managed populations: simple models and an application to the Islay Barnacle Goose population. Dissertation. Department of Statistics and Modelling Science. University of Strathclyde, Glasgow, Scotland (cit. Murdock 1994).

Miller, R. S. (1967). *Pattern and Process in Competition.* Advances in Ecological Research 4, London, Academic Press, pp. 1–74.

Molofsky, J. and Bever, J. D. (2004). A new kind of ecology? *BioScience* **54**, 440–446.

Moloney, K. A. and Levin, S. A. (1996). The effects of disturbance architecture on landscape-level population dynamics. *Ecology* **77**, 375–394.

Mooers, A. O. and Harvey, P. H. (1994). Metabolic rate, generation time and the rate of molecular evolution in birds. *Molecular Phylogeny and Evolution* **3**, 344–350.

Mora, C., Chittaro, P. M., Sale, P. F., Kritzer, J. P. and Ludsin, S. A. (2003). Patterns and processes in reef fish diversity. *Nature* **42**, 933–936.

Morand, S., Poulin, R., Rohde, K. and Hayward, C. J. (1999). Aggregation and species coexistence of ectoparasites of marine fishes. *International Journal for Parasitology* **29**, 663–672.

Morand, S., Rohde, K. and Hayward, C. J. (2002). Order in parasite communities of marine fish is explained by epidemiological processes. *Parasitology* **124**, S57–S63.

Mouillot, D., George-Nascimento, M. and Poulin, R. (2003), How parasites divide resources: a test of the niche apportionment hypothesis. *Journal of Animal Ecology* **72**, 757–764.

Moulton, M. P. and Pimm, S. L. (1987). Morphological assortment in introduced Hawaiian passerines. *Evolutionary Ecology* **1**, 113–124.

Moyle, P. B. and Herbold, B. (1987). Life-history patterns and community structure in stream fishes of western North America: comparisons with eastern North America and Europe. In Matthews, W. J, and Heins, D. C., eds., *Community and Evolutionary Ecology of North American Stream Fishes*. Norman, University of Oklahoma Press, pp. 25–32.

Murdoch, W. W. (1994). Population regulation in theory and practice. *Ecology* **75**, 271–287.

Myers, N. (1997). The niche diversity of biodiversity issues. In Reaka-Kudla, M. L., Wilson, D. E. and Wilson, E. U., eds., *Biodiversity II. Understanding and Perfecting our Biological Resources*. Washington, Joseph Henry Press, pp. 125–138.

Neubert, M. G. (1997). A simple population model with qualitatively uncertain dynamics. *Journal of Theoretical Biology* **189**, 399–411

Nichols, S. P. (1996). Nematodes of rabbits and hares of eastern Australia. BSc. Honours thesis, University of New England, Armidale, Australia.

Nicholson, A. J. (1933). The balance of animal populations. *Journal of Animal Ecology* **2**, 132–178.

(1954). An outline of the dynamics of animal populations. *Australian Journal of Zoology* **2**, 9–65.

Nisbet, R. M. and Gurney, W. S. C. (1982). *Modelling Fluctuating Populations*. Chichester, Wiley.

Nollen, P. M. (1993). *Echinostoma trivolvis*: mating behavior of adults raised in hamsters. *Parasitology Research* **79**, 130–132.

Norton, J., Lewis, J. W. and Rollinson, D. (2003). Parasite infracommunity diversity in eels: a reflection of local component community diversity. *Parasitology* **127**, 475–482.

Novotny, V., Basset, Y., Miller, S. E. *et al.* (2002). Low host specificity of herbivorous insects in a tropical forest. *Nature* **416**, 841–844.

Nowak, M. A. and May, R. M. (1994). Superinfection, metapopulation dynamics, and the evolution of diversity. *Journal of Theoretical Biology* **170**, 95–114.

Oberdorff, T. B., Hugueny, A., Compin, A. and Belkessam, D. (1998). Non-interactive fish communities in the coastal streams of North-western France. *Journal of Animal Ecology* **67**, 472–484.

Orians, G. H. (1962). Natural selection and ecological theory. *American Naturalist* **46**, 257–263.

Ovaskainen, O. and Hanski, I. (2002). Transient dynamics in metapopulation response to perturbation. *Theoretical Population Biology* **61**, 285–295.

Paine, R. T. (2002). Advances in ecological understanding: by Kuhnian revolution or conceptual evolution? *Ecology* **83**, 1553–1559.

Palmer, M. W. (2001). Extending the quasi-neutral concept. *Folia Geobotanica* **36**, 25–33.

Pandian, T. J. and Vivekanadan, E. (1985). Energetics of feeding and digestion. In Tyler, P. and Calow, P., eds., *Fish Energetics – New Perspectives*. London, Croom Helm, pp. 99–124.

Park, T. (1954). Experimental studies of interspecies competition. II. Temperature, humidity, and competition in two species of *Tribolium*. *Physiological Zoology* **27**, 177–238.

Pence, D. B. (1990). Helminth communities of mammalian hosts: concepts at the infracommunity, component and compound community. In Esch, G., Bush, A. and Aho, J., eds., *Parasite Communities: Patterns and Processes*. London, New York, Chapman and Hall, pp. 233–260.

Petraitis, P. S., Latham, R. E. and Niesenbaum, R. A. (1989). The maintenance of species diversity by disturbance. *Quarterly Review of Biology* **64**, 393–418.

Pianka, E. R. (1973). The structure of lizard communities. *Annual Review of Ecology and Systematics* **4**, 53–74.

(1974). *Evolutionary Ecology*. 2nd edn. New York, Harper and Row.

(1983). *Evolutionary Ecology*. 3rd edn. New York, Harper and Row.

Pickett, S. T. A. (1980). Non-equilibrium coexistence of plants. *Bulletin of the Torrey Botanical Club* **107**, 238–248.

Pielou, E. C. (1969). *An Introduction to Mathematical Ecology*. New York, Wiley-Interscience.

(1975). *Ecological Diversity*. New York, Wiley-Interscience.

(1979). *Biogeography*. New York, Wiley-Interscience.

Pimm, S. L. (1978). An experimental approach to the effects of predictability on community structure. *American Zoologist* **18**, 797–808.

(1991). *The Balance of Nature?* Chicago and London, The University of Chicago Press.

Poulin, R. (1998). *Evolutionary Ecology of Parasites*. London, Chapman and Hall.

Poulin, R. and Mouillot, D. (2003). Host introductions and the geography of parasite taxonomic diversity. *Journal of Biogeography* **30**, 1–9.

Poulin, R. and Rohde, K. (1997). Comparing the richness of metazoan ectoparasite communities of marine fishes: controlling for host phylogeny. *Oecologia* **110**, 278–283.

Poulin, R., Mouillot, D. and George-Nascimento, M. (2003). The relationship between species richness and productivity in parasite communities. *Oecologia* **137**, 277–285.

Price, P. W. (1977). General concepts of the evolutionary biology of parasites. *Evolution* **31**, 405–420.

(1980). *Evolutionary Biology of Parasites*. Princeton, N. J., Princeton University Press.

(1983). Communities of specialists: Vacant niches in ecological and evolutionary time. In Strong, D., Simberloff, D. and Abele, L., eds., *Ecological Communities: Conceptual Issues and the Evidence*. Princeton, N. J., Princeton University Press.

(1984). Alternative paradigms in community ecology. In Price, P. W., Slobodchikoff, C. N. and Gaud, W. S., eds., *A New Ecology. Novel Approaches to Interactive Systems*. New York, Chichester, Brisbane, Toronto, Singapore, John Wiley & Sons, pp. 353–383.

Price, P. W., Gaud, W. S, and Slobodchikoff, C. N. (1984). Introduction: is there a new ecology? In Price, P. W., Slobodchikoff, C. N. and Gaud, W. S., eds.,

A New Ecology. Novel Approaches to Interactive Systems. New York, John Wiley & Sons, pp. 1–11.

Price, P. W., Slobodchikoff, C. N. and Gaud, W. S., eds., (1984). *A New Ecology. Novel Approaches to Interactive Systems.* New York, John Wiley & Sons.

Priddle, D. (1987). The mobility and habitat utilisation of kangaroos. In Caughley, G., Shepherd, N. and Short, J., eds., *Kangaroos: Their Ecology and Management in the Sheep Rangelands of Australia.* Cambridge, Cambridge University Press, pp. 100–118.

Pulliam, H. R. and Enders, F. A. (1971). The feeding ecology of five sympatric finch species. *Ecology* **52**, 557–566.

Rahbeck, C. (1993). Captive breeding – a useful tool in the preservation of biodiversity? *Biodiversity and Conservation* **2**, 428–437.

Rahbeck, C. and Graves, G. R. (2001). Multiscale assessment of patterns of avian species richness. *Proceedings of the National Academy of Science USA* **98**, 4534–4539.

Ramasamy, P., Ramalingam, K., Hanna, R. E. B. and Halton, D. W. (1985). Microhabitats of gill parasites (Monogenea and Copepoda of teleosts (*Scomberoides* spp.)). *International Journal for Parasitology* **15**, 385–397.

Ramensky, L. G. (1926). Die Grundgesetzmässigkeiten im Aufbau der Vegetationsdecke. *Botanisches Centralblatt, N.F.* **7**, 453–455.

Rathcke, B. J. (1976a). Insect plant patterns and relationships in the stem-boring guild. *American Midland Naturalist* **99**, 98–117.

(1976b). Competition and coexistence within a guild of herbivorous insects. *Ecology* **57**, 76–87.

Rensch, B. (1954). *Neuere Probleme der Abstammungslehre.* Ferdinand Enke Verlag, Stuttgart.

Ricklefs, R. E. (2004). A comprehensive framework for global patterns in biodiversity. *Ecology Letters* **7**, 11–15.

Ricklefs, R. E. and Miller, G. L. (1999). *Ecology.* 4th edn. New York, W. H. Freeman and Company.

Ritchie, M. E. and Olff, H. (1999). Spatial scaling laws yield a synthetic theory of biodiversity. *Nature* **400**, 557–562.

Robertson, D. R. (1991). Increases in surgeonfish populations after mass mortality of the sea urchin *Diadema antillarum* in Panama indicate food limitation. *Marine Biology* **11**, 437–444.

(1995). Competitive ability and the potential for lotteries among territorial reef fishes. *Oecologia* **103**, 180–190.

(1996). Interspecific competition controls abundance and habitat use of territorial Caribbean damselfishes. *Ecology* **77**, 885–899.

(2001). Population maintenance among tropical reef fishes: inferences from small-island endemics. *Proceedings of the National Academy of Sciences* **98**, 5667–5670.

Robertson, D. R. and Gaines, S. D. (1986). Interference competition structures habitat use in a local assemblage of coral reef surgeonfishes. *Ecology* **67**, 1372–1383.

Robertson, G., Short, J. and Wellard, G. (1987). The environment of the Australian sheep rangelands. In *Kangaroos: Their Ecology and Management in the Sheep Rangelands of Australia.* Cambridge, Cambridge University Press, pp. 14–34.

Rohde, K. (1968). Die Entwicklung von *Multicotyle purvisi*. *Zeitschrift für Parasitenkunde* **30**, 278–280.

(1972). The Aspidogastrea, especially *Multicotyle purvisi* Dawes 1941. *Advances in Parasitology* **10**, 77–151.

(1973). Structure and development of *Lobatostoma manteri* sp.nov. (Trematoda, Aspidogastrea) from the Great Barrier Reef, Australia. *Parasitology* **66**, 63–83.

(1975). Early development and pathogenesis of *Lobatostoma manteri* Rohde (Trematoda: Aspidogastrea). *International Journal for Parasitology* **5**, 597–607.

(1976a). Marine parasitology in Australia. *Search* **7**, 477–482.

(1976b). Monogenean gill parasites of *Scomberomorus commersoni* Lacépède and other mackerel on the Australian east coast. *Zeitschrift für Parasitenkunde* **51**, 49–69.

(1977a). Species diversity of monogenean gill parasites of fish on the Great Barrier Reef. Proceedings of the Third International Coral Reef Symposium Miami-Florida, pp. 585–591.

(1977b). A non-competitive mechanism responsible for restricting niches. *Zoologischer Anzeiger* **199**, 164–172.

(1977c). Habitat partitioning in Monogenea of marine fishes. *Heteromicrocotyla australiensis*, sp. nov. and *Heteromicrocotyloides mirabilis*, gen. and sp. nov. (Heteromicrocotylidae) on the gills of *Carangoides emburyi* (Carangidae) on the Great Barrier Reef, Australia. *Zeitschrift für Parasitenkunde* **53**, 171–182.

(1978a). Latitudinal differences in species diversity and their causes. I. A review of the hypotheses explaining the gradients. *Biologisches Zentralblatt* **97**, 393–403.

(1978b). Latitudinal gradients in species diversity and their causes. II. Marine parasitological evidence for a time hypothesis. *Biologisches Zentralblatt* **97**, 405–418.

(1978c). Latitudinal differences in host specificity of marine Monogenea and Digenea. *Marine Biology* **47**, 125–134.

(1979a). A critical evaluation of intrinsic and extrinsic factors responsible for niche restriction in parasites. *American Naturalist* **114**, 648–671.

(1979b). The buccal organ of some Monogenea Popyopisthocotylea. *Zoologica Scripta* **8**, 161–170.

(1980a). Warum sind ökologische Nischen begrenzt? Zwischenartlicher Antagonismus oder innerartlicher Zusammenhalt? *Naturwissenschaftliche Rundschau* **33**, 98–102.

(1980b). Comparative studies on microhabitat utilization by ectoparasites of some marine fishes from the North Sea and Papua New Guinea. *Zoologischer Anzeiger* **204**, 27–63.

(1980c). Diversity gradients of marine Monogenea in the Atlantic and Pacific Oceans. *Experientia* **36**, 1368–1369.

(1980d). Species diversification, with special reference to parasites. Proceedings of the 24th Conference of the Australian Society for Parasitology, Adelaide.

(1980e). Host specificity indices of parasites and their application. *Experientia* **36**, 1369–1371.

(1981a). Niche width of parasites in species-rich and species-poor communities. *Experientia*, **37**, 359–361.

(1981b). Population dynamics of two snail species, *Planaxis sulcatus* and *Cerithium moniliferum*, and their trematode species at Heron island, Great Barrier Reef. *Oecologia* **49**, 344–352.

(1982). *Ecology of Marine Parasites*. St. Lucia. Brisbane, University of Queensland Press.

(1984). Ecology of marine parasites. In Kinne, O. and Bulnheim, H. P., eds., *Diseases of Marine Organisms*, Helgoländer Meeresuntersuchungen **37**, 5–33.

(1985). Increased viviparity of marine parasites at high latitudes. *Hydrobiologia* **127**, 197–201.

(1986). Differences in species diversity of Monogenea between the Pacific and Atlantic Oceans. *Hydrobiologia* **137**, 21–28.

(1989). Simple ecological systems, simple solutions to complex problems? *Evolutionary Theory* **8**, 305–350.

(1991). Intra- and interspecific interactions in low density populations in resource-rich habitats. *Oikos* **60**, 91–104.

(1992). Latitudinal gradients in species diversity: the search for the primary cause. *Oikos* **65**, 514–527.

(1993). *Ecology of Marine Parasites*. 2nd edn. Wallingford, UK, CAB – International (Commonwealth Bureaux of Agriculture).

(1994a). Niche restriction in parasites: proximate and ultimate causes. *Parasitology* **109**, S69–S84.

(1994b). The minor groups of parasitic Platyhelminthes. *Advances in Parasitology* **33**, 145–234.

(1997). The larger area in the tropics does not explain latitudinal gradients in species diversity. *Oikos* **79**, 169–172.

(1998a). Latitudinal gradients in species diversity. Area matters, but how much? *Oikos* **82**, 184–190.

(1998b). Is there a fixed number of niches for endoparasites of fish? *International Journal for Parasitology* **28**, 1861–1865.

(1999). Latitudinal gradients in species diversity and Rapoport's rule revisited: a review of recent work, and what can parasites teach us about the causes of the gradients? *Ecography*, ***22***, 593–613 (invited Minireview on the occasion of the 50th anniversary of the Nordic Ecological Society Oikos). Also published in Fenchel, T., ed., *Ecology 1999 and Tomorrow*. University of Lund, Sweden, Oikos Editorial Office, pp. 73–93.

(2001a). Spatial scaling laws may not apply to most animal species. *Oikos* **93**, 499–504

(2001b). The Aspidogastrea: an archaic group of Platyhelminthes. In Littlewood, D. T. J. and Bray, R. A., eds., *Interrelationships of the Platyhelminthes*. London and New York, Taylor and Francis, pp. 159–167.

(2001c). Parasitism. In Levin, S. ed., *Encyclopedia of Biodiversity* Vol. I. New York, Academic Press, pp. 463–484.

(2002). Ecology and biogeography of marine parasites. *Advances in Marine Biology* **43**, 1–86.

(2005a). Cellular automata and ecology. *Oikos* **110**, 203–207.

(2005b). Eine neue Ökologie. Aktuelle Probleme der evolutionären Ökologie. *Naturwissen schaftliche Rundschau* (in press).

Rohde, K. and Hayward, C. J. (2000). Oceanic barriers as indicated by scombrid fishes and their parasites. *International Journal for Parasitology* **30**, 579–583.

Rohde, K. and Heap, M. (1998). Latitudinal differences in species and community richness and in community structure of metazoan endo- and ectoparasites of marine teleost fish. *International Journal for Parasitology* **28**, 461–474.

Rohde, K. and Hobbs, R. (1986). Species segregation: Competition or reinforcement of reproductive barriers? In Cremin, M., Dobson, C. and Moorhouse, D. E., eds., *Parasite lives. Papers on Parasites, their Hosts and their Association to Honour J. F. A. Sprent.* St. Lucia, London, New York, University of Queensland Press, pp. 89–199.

(1988) Rarity in marine Monogenea. Does an Allee-effect or parasite-induced mortality explain truncated frequency distributions? *Biologisches Zentralblatt* **107**, 327–338.

(1999). An asymmetric percent similarity index. *Oikos* **87**, 601–602.

Rohde, K. and Rohde, P. P. (2001). Fuzzy chaos: reduced chaos in the combined dynamics of several independently chaotic populations. *American Naturalist* **158**, 553–556.

(2005). The ecological niches of parasites. In: Rohde, K. ed., *Marine Parasitology.* CSIRO Publishing, Melbourne (in press).

Rohde, K., Hayward, C. Heap, M. and Gosper, D. (1994). A tropical assemblage of ectoparasites: gill and head parasites of *Lethrinus miniatus* (Teleostei, Lethrinidae). *International Journal for Parasitology* **24**, 1031–1053.

Rohde, K., Hayward, C. and Heap, M. (1995). Aspects of the ecology of metazoan ectoparasites of marine fishes. *International Journal for Parasitology* **25**, 945–970.

Rohde, K., Roubal, F. and Hewitt, G. C. (1980). Ectoparasitic Monogenea, Digenea, and Copepoda from the gills of some marine fishes of New Caledonia and New Zealand. *New Zealand Journal of Marine and Freshwater Research* **14**, 1–13.

Rohde, K., Worthen, W., Heap, M., Hugueny, B. and Guégan, J. F. (1998). Nestedness in assemblages of metazoan ecto- and endoparasites of marine fish. *International Journal for Parasitology* **28**, 543–549.

Rosenheim, J. A. and Tabashnik, B. E. (1993). Generation time and evolution. *Nature* **365**, 791–792.

Rosenzweig, M. L. (1973). Evolution of the predator isocline. *Evolution* **27**, 89–94.

(1995). *Species Diversity in Space and Time.* Cambridge, Cambridge University Press.

Rosenzweig, M. L. and Sandlin, E. A. (1997). Species diversity and latitude: listening to area's signal. *Oikos* **80**, 172–176.

Rosenzweig, M. L. and Sterner, P. W. (1970). Population ecology of desert rodent communities: body size and seed-husking as bases for heteromyid coexistence. *Ecology* **51**, 217–224.

Rosenzweig, M. L. and Ziv, Y. (1999). The echo pattern of species diversity: pattern and processes. *Oikos* **22**, 614–628.

Roubal, F. R. (1979). The taxonomy and site specificity of the ectoparasitic metazoans on the black bream, *Acanthopagrus australis* (Günther), in northern New South Wales. *Australian Journal of Zoology Supplement* **84**.

Roughgarden, J. (1989). The structure and assembly of communities. In Roughgarden, J. S. D., May, R. M. and Levin, S. A., eds., *Perspectives in Ecological Theory.* Princeton, N. J., Princeton University Press, pp. 203–226.

Roy, K., Jablonski, D., Valentine, J. W. *et al.* (1998). Marine latitudinal diversity gradients: tests of causal hypotheses. *Proceedings of the National Academy of Sciences USA* **95**, 3699–3702.

Royama, T. (1981) Fundamental concepts and methodology for the analysis of animal population dynamics, with special reference to univoltine species. *Ecological Monographs* **51**, 473–493.

Sale, P. F. (1977). Maintenance of high diversity in coral reef fish communities. *American Naturalist* **111**, 337–359.

(1991). Reef fish communities: open nonequilibrial systems. In Sale, P. F., ed., *The Ecology of Fishes on Coral Reefs*. San Diego, Academic Press, pp. 564–598.

(2002). The science we need to develop more effective management. In Sale, P. F., ed., *Coral Reef Fishes. Dynamics and Diversity in a Complex Ecosystem*. Amsterdam, Academic Press, pp. 361–376.

Sale, P. F., ed. (2002). *Coral reef fishes. Dynamics and Diversity in a Complex Ecosystem*. Amsterdam, Academic Press.

Sale, P. F. and Tolimieri, N. (2000). Density-dependence at some time and place? *Oecologia* **124**, 166–171.

Sale, P. F., Doherty, P. J., Eckert, G. J., Douglas, W. A. and Ferrell, D. J. (1984) Large scale spatial and temporal variation in recruitment to fish populations on coral reefs. *Oecologia* **64**, 191–198.

Schad, G. A. (1962). Gause's hypothesis in relation to the oxyuroid populations of *Testudo graeca*. *Journal of Parasitology* **48**, 36.

(1963). Niche diversification in a parasitic species flock. *Nature* **198**, 404–406.

Schoener, T. W. (1983). Field experiments on interspecific competition. *American Naturalist* **122**, 240–285.

(1986a). Overview: kinds of ecological communities-ecology becomes pluralistic. In Diamond, J. and Case, T. J., eds., *Community Ecology*. New York, Harper and Row, pp. 467–479.

(1986b). Resource partitioning. In Kikkawa, J. and Anderson, D. J., eds., *Community Ecology: Pattern and Process*. Melbourne, Blackwell Scientific Publishing, pp. 91–126.

Shepherd, J. G. and Cushing, D. H. (1990). Regulation in fish populations: myth or mirage In Hassell, M. P. and May, R. M., eds., *Population, Regulation and Dynamics. Philosophical Transactions of the Royal Society of London, Series B* **330**, 151–164.

Short, J. (1987). Factors affecting food intake of rangeland herbivores. In Caughley, G., Shepherd, N. and Short, J., eds., *Kangaroos: Their Ecology and Management in the Sheep Rangelands of Australia*. Cambridge, Cambridge University Press, pp. 84–99.

Shurin, J. B. (2000). Dispersal limitation, invasion resistance, and the structure of pond zooplankton communities. *Ecology* **81**, 3074–3086.

Shurin, J. B. and Srivastava, D. S. (in press) New perspectives on local and regional diversity: Beyond saturation. In Holyoak, M., Holt, R. and Leibold, M. eds., *Metacommunities*. Chicago, IL, University of Chicago Press.

Shurin, J. B., Havel, J. E., Leibold, M. A. and Pinel-Alloul, B. (2000). Local and regional zooplankton species richness: a scale-independent test for saturation. *Ecology* **81**, 3062–3072.

Sibley, C. G. and Ahlquist, , J. E. (1990). *Phylogeny and Classification of Birds. A Study in Molecular Evolution*. Yale, Yale University Press.

Silman, M. R., Terborgh, J. W. and Kiltie, R. A. (2003). Population regulation of a dominant rain forest tree by a major seed predator. *Ecology* **84**, 431–438.

Silverton, J. (1980). The dynamics of a grassland ecosystem: botanical equilibrium in the park grassland experiment. *Journal of Applied Ecology* **17**, 491–504.

Silverton, J. and Charlesworth, D. (2001). *Introduction to Plant Population Biology*. 4th edn. Oxford, Blackwell Science.

Simberloff, D. S. (1974). Equilibrium theory of island biogeography and ecology. *Annual Review of Ecology and Systematics* **5**, 161–182.

(1976). Species turnover and equilibrium island biogeography. *Science* **194**, 572–578.

(1978). Using island biogeographic distributions to determine if colonisation is stochastic. *American Naturalist* **112**, 713–726.

Simberloff, D. S. and Wilson, E. O. (1970). Experimental zoogeography of islands. A two-year record of colonization. *Ecology* **51**, 934–937.

Simkova, A., Desdevises, Y., Gelnar, M. and Morand, S. (2000). Co-existence of nine gill ectoparasites (*Dactylogyus*: Monogenea) parasitising the roach *Rutilus rutilus* (L.): history and present ecology. *International Journal for Parasitology* **30**, 1077–1088.

Simkova, A., Gelnar, M. and Morand, S. (2001a). Order and disorder in ectoparasite communities: the case of congeneric gill monogeneans (*Dactylogyrus* spp.). *International Journal for Parasitology* **31**, 1205–1210.

Simkova, A., Gelnar, M. and Sasal, P. (2001b). Aggregation of congeneric parasites (Monogenea: *Dactylogyrus*). *Parasitology* **123**, 599–607.

Simkova, A., Desdevises, Y., Gelnar, M. and Morand, S. (2001c). Morphometric correlates of host specificity in *Dactylogyrus* species (Monogenea) parasites of European Cyprinid fish. *Parasitology* **123**, 169–177.

Simkova, A., Ondrackova, M., Gelnar, M. and Morand, S. (2002). Morphology and coexistence of congeneric ectoparasite species: reinforcement of reproductive isolation? *Biological Journal of the Linnaean Society* **76**, 125–135.

Sinclair, A. R. E. (1979). The eruption of the ruminants. In Sinclair, A. R. E. and Norton-Griffiths, M., eds., *Serengeti – the Dynamics of an Ecosystem*. Chicago, University of Chicago Press, pp. 82–103.

(1985). Does interspecific competition or predation shape the African ruminant community? *Journal of Animal Ecology* **54**, 899–918.

Smith, F. D. M., May, R. M., Pellew, T. H., Johnson, T. H. and Walter, K. R. (1993). How much do we know about the current extinction rate? *Trends in Ecology and Evolution* **8**, 375–378.

Sousa, W. P. (1990). Spatial scale and the processes structuring a guild of larval trematode parasites. In Esch, G., Bush, A. O. and Aho, J. M., eds., *Parasite Communities: Patterns and Processes*. London, New York, Chapman and Hall, pp. 41–67.

(1992). Interspecific interactions among larval trematode parasites of freshwater and marine snails. *American Zoologist* **32**, 583–592.

(1993). Interspecific antagonism and species coexistence in a diverse guild of larval trematode parasites. *Ecological Monographs* **63**, 103–128.

Srivastava, D. S. (1999). Using local-regional richness plots to test for species saturation: pitfalls and potentials. *Journal of Animal Ecology* **68**, 1–16.

Srivastava, D. S., Lawton, J. H. and Robinson, G. S. (1997). Spore-feeding: a new, regionally vacant niche for bracken herbivores. *Ecological Entomology* **22**, 475–478.

Stauffer, D. and Chowdhury, D. (2005). Evolutionary ecology in-silico: evolving food webs, migrating population and speciation. *Physica A* **352**, 202–215.

Steadman, D. W. (1995). Prehistoric extinctions of Pacific island birds. Biodiversity meets zooarchaeology. *Science* **267**, 1123–1131.

Stehli, F. G., Douglas, D. G. and Newell, N. D. (1969). Generation and maintenance of gradients in taxonomic diversity. *Science* **164**, 947–949.

Stevens, G. C. (1989). The latitudinal gradients in geographical range: how so many species co-exist in the tropics. *American Naturalist* **133**, 240–256.

Stiassny, M. L. J. and Raminosoa, N. (1994). The fishes of the inland waters of Madagascar. In Teugels, G. G., Guegan, J. F. and Albert, J. J., eds., *Biological Diversity in African Fresh- and Brackish Water Fishes*. Tervuren, Belgique, Annales Musee Royal de l'Afrique Centrale, pp. 133–149.

Strong, D. R., Jr. (1981). The possibility of insect communities without competition.: Hispine beetles on *Heliconia*. In Denno, R. F. and Dingle, H., eds., *Insect Life History Patterns: Habitat and Geographic Variation*, New York, Springer-Verlag, pp. 183–194.

(1984). Density-vague ecology and liberal population regulation in insects. In Price, P. W., Slobodchikoff, C. N. and Gaud, W. S., eds., *A New Ecology. Novel Approaches to Interactive Systems*. New York, Chichester, Brisbane, Toronto, Singapore, John Wiley & Sons, pp. 313–327.

Strong, D. R. Jr., Szyska, L. A. and Simberloff, D. S. (1979). Tests of community-wide character displacement against null hypotheses. *Evolution* **33**, 897–913.

Strong, D. R., Jr., Lawton, J. H. and Southwood, T. R. E. (1984). *Insects on Plants. Community Patterns and Mechanisms*. Oxford, Blackwell Scientific.

Sugihara, G., Grenfell, B. and May, R. M. (1990). Distinguishing error from chaos in ecological time. In Hassell, M. P. and May, R. M., eds., *Population, Regulation and Dynamics. Philosophical Transactions of the Royal Society of London, Series B* **330**, 235–251.

Surman, C. A. and Wooller, R. D. (2003). Comparative foraging ecology of five sympatric terns at a sub-tropical island in the eastern Indian Ocean. *Journal of Zoology (London)* **259**, 219–230.

Tarr, H. L. A. (1969). Contrast between fish and warm blooded vetebrates in enzyme systems of intermediary metabolism. In Neuhaus, O. W. and Halves, J. E., eds., *Fish in Research*. New York, Academic Press, pp. 155–174.

Tauber, C. A. (1978). Response to criticism of Hendrickson. *Science* **200**, 346.

Tauber, C. A. and Tauber, M. J. (1977). Sympatric speciation based on allelic changes in three loci: evidence from natural populations in two habitats. *Science* **197**, 1298–1299.

(1987). Food specificity in predaceous insects: a comparative ecophysiological and genetic study. *Evolutionary Ecology* **1**, 175–186.

Terborgh, J. (1973). On the notion of favourableness in plant ecology. *American Naturalist* **107**, 481–501.

Terborgh, J. and Faaborg, J. (1980). Saturation of bird communities in the West Indies. *American Naturalist* **116**, 178–195.

Terborgh, J. and Weske, J. S. (1969). Colonization of secondary habitats by Peruvian birds. *Ecology* **50**, 765–782.

Tilman, D. (1982). *Resource Competition and Community Structure*. Princeton, N. J., Princeton University Press.

(1999). Diversity by default. *Science* **283**, 495–496.

Tilman, D. and Kareiva, P. (1997). *Spatial Ecology: The Role of Space in Population Dynamics and Interspecific Interactions*. Monographs in Population Biology, no. 30. Princeton, N. J., Princeton University Press.

Tilman, D., Lehman, C. L. and Kareiva, P. (1997). Population dynamics in spatial habitats. In Tilman, D. and Kareiva, P., eds., *Spatial Ecology: the Role of Space in Population Dynamics and Interspecific Interactions*. Monographs in Population Biology, no. 30. Princeton, N. J., Princeton University Press, pp. 3–20.

Timoféeff-Ressovsky, N. W., Zimmer, K. G. and Delbrück, M. (1935). Über die Natur der Genmutation und der Genstruktur. *Nachrichten aus der Biologie der Gesellschaft der Wissenschaften Göttingen* **1**, 189–245.

Tokeshi, M. (1990). Niche apportionment or random assortment: species abundance patterns. *Journal of Animal Ecology* **59**, 1129–1146.

(1999). *Species Coexistence: Ecological and Evolutionary Perspectives*. Oxford, Blackwell Science.

Tonn, W. M., Magnuson, J. J., Rask, M. and Toivonen, J. (1990). Intercontinental comparison of small-lake fish assemblages: the balance between local and regional processes. *American Naturalist* **136**, 345–375.

Torchin M. and Kuris A. (2005). Introduced parasites and the use of parasites for controlling introduced pest species. In Rohde, K., ed. *Marine Parasites. An Introduction*, Melbourne, CSIRO Publishing.

Torchin, M. E., Lafferty, K. D., Dobson, A. P., McKenzie, V. J. and Kuris, A. M. (2003). Introduced species and their missing parasites. *Nature* **421**, 628–630.

Tsukamoto, K. and Aoyama, J. (1998). Evolution of freshwater eels of the genus *Anguilla*: a probable scenario. *Environmental Biology of Fishes* **52**, 139–148.

Tsukamoto, K., Aoyama, J. and Miller, M. J. (2002) Migration, speciation, and the evolution of diadromy in anguillid eels. *Canadian Journal of Fisheries and Aquatic Science* **59**, 1989–1998.

Turchin, P. (1995). Population regulation: old arguments and a new synthesis. In Cappuccino, N. and Price, P. W., eds., *Population Dynamics: New Approaches and Synthesis*. San Diego, Academic Press, pp. 19–41.

Tyler, H., Brown, K. S. and Wilson, K. (1994). *Swallowtail Butterflies of the Americas*. Gainsville, Scientific Publishing.

Underwood, T. (1986). The analysis of competition by field experiments. In Kikkawa, J. and Anderson, D. J., eds., *Community Ecology: Pattern and Process*. Melbourne, Blackwell Scientific Publishing, pp. 240–268.

Valdovinos, C., Navarete, S. A. and Marquet, P. A. (2003). Mollusc species diversity in the Southeastern Pacific: why are there more species towards the pole? *Ecography* **26**, 129–144.

Van Nouhuys, S. and Hanski, I. (2002). Colonization rates and distances of a host butterfly and two specific parasitoids in a fragmented landscape. *Journal of Animal Ecology* **71**, 639–650.

Van Valen, L. (1973). A new evolutionary law. *Evolutionary Theory* **1**, 1–30.

Varley, G. C. (1947). The natural control of population balance in the knapweed gall-fly (*Urophora jaceana*). *Journal of Animal Ecology* **16**, 139–187.

Via, S. (2001). Sympatric speciation in animals: the ugly duckling grows up. *Trends in Ecology and Evolution* **16**, 381–390.

Volkov, I., Banavar, J. R., Hubbell, S. P. and Maritan, A. (2003). Neutral theory and relative species abundance in ecology. *Nature* **424**, 1035–1037.

Von Bertalanffy, L. (1952). *Problems of Life*. London, Watts and Co.

(1973). *General System Theory*. Harmondsworth, Penguin University Books.

Waide, R. B., Willig, M. R., Steiner, C. F., Mittelbach, G. G., Gough, L., Dodson, S. I., Juday, G. P. and Parmenter, R. (1999). The relationship between productivity and species richness. *Annual Review of Ecology and Systematics* **30**, 257–300.

Walker, J. C. (1979). *Austrobilharzia terrigalensis*: a schistosome dominant in interspecific interactions in a molluscan host. *International Journal for Parasitology* **9**, 137–140.

Walker, T. D. and Valentine, J. W. (1984). Equilibrium models of evolutionary diversity and the number of empty niches. *American Naturalist* **124**, 887–899.

Walter, G. H. (1995). Species concepts and the nature of ecological generalizations about diversity. In Lambert, D. M. and Spencer, H. G., eds., *Speciation and the Recognition Concept*. Baltimore and London, John Hopkins University Press, pp. 191–224.

Walter, G. H. and Hengeveld, R. (2000). The structure of the two ecological paradigms. *Acta Biotheoretica* **48**, 15–46.

Walter, G. H. and Patterson, H. E. H. (1994). The implications of palaeontological evidence for theories of ecological communities and species richness. *Australian Journal of Ecology* **19**, 241–250.

(1995). Levels of understanding in ecology: interspecific competition and community ecology. *Australian Journal of Ecology* **20**, 463–466.

Watkinson, A. R. (1997). Plant population dynamics. In Crawley, M. J., ed., *Plant Ecology*, 2nd edn. Oxford, Blackwell Science, pp. 359–400.

Webster, M. S. (2003). Temporal density dependence and population regulation in a marine fish. *Ecology* **84**, 623–628.

Werner, R. R. and Hughes, T. P. (1988). The population dynamics of reef fishes. *Proceedings of the 6th International Coral Reef Symposium* **I**, 144–155.

Wetzel, R. G. (1975). *Limnology*. Philadelphia, PA, Saunders.

White, T. C. R. (1993). *The Inadequate Environment: Nitrogen and the Abundance of Animals*. Berlin, Springer-Verlag.

Whittaker, R. H. (1967). Gradient analysis of vegetation. *Biological Review* **42**, 207–264.

(1972). Evolution and measurement of species diversity. *Taxon* **21**, 213–251.

(1975). *Communities and Ecosystems*. New York, Macmillan.

Wiens, J. A. (1974). Habitat heterogeneity and avian community structure in North American grasslands. *American Midland Naturalist* **91**, 195–213.

(1984). Resource systems, populations, and communities. In Price, P. W., Slobodchikoff, C. N. and Gaud, W. S., eds., *A New Ecology. Novel Approaches to Interactive Systems*. New York, Chichester, Brisbane, Toronto, Singapore, John Wiley & Sons, pp. 397–436.

Willig, M. R. (2001). Latitude, common trends within. In Levin, S., ed., *Encyclopedia of Biodiversity*, vol.3. New York, Academic Press, pp. 701–714.

Willig, M. R., Kaufman, D. M. and Stevens, R. D. (2003). Latitudinal gradients of biodiversity: pattern, process, scale, and synthesis. *Annual Review of Ecology, Evolution and Systematics* **34**, 273–309.

Wilson, D. S. (1975). The adequacy of body size as niche difference. *American Naturalist* **109**, 769–784.

(1969). The species equilibrium. In Woodwell, G. M. and Smith, H. H., eds., *Diversity and Stability in Ecological Systems. Brookhaven Symposia in Biology, no.22.* Upton, New York, Brookhaven National Laboratory, pp. 38–47.

Wilson, E. O. and Simberloff, D. S. (1969). Experimental zoogeography of islands. Defaunation and monitoring techniques. *Ecology* **50**, 267–277.

Wolfram, S. (1986). *Theory and Applications of Cellular Automata: Advanced Series on Complex Systems.* Singapore, World Scientific Publishing.

(2002). *A New Kind of Science.* Champaign, Il., Wolfram Media Inc.

Worthen, W. B. and Rohde, K. (1996). Nested subset analysis of colonisation-dominated communities: metazoan ectoparasites of marine fish. *Oikos* **75**, 471–478.

Wright, S. D., Gray, R. D. and Gardner, R. C. (2003). Energy and the rate of evolution: inferences from plant rDNA substitution rates in the Western Pacific. *Evolution* **57**, 2893–2898.

Wright, S. J. (2002). Plant diversity in tropical forests: a review of mechanisms of species coexistence. *Oecologia* **130**, 1–14.

Wynne-Edwards, V. C. (1962). *Animal Dispersion in Relation to Social Behaviour.* Edinburgh, Oliver and Boyd.

Zhang, H. and Wu, J. (2002). A statistical thermodynamic model of the organizational order of vegetation. *Ecological Modelling* **153**, 69–80.

Zwölfer, H. (1974a). Innerartliche Kommunikationssysteme bei Bohrfliegen. *Biologie in unserer Zeit* **6**, 147–153.

(1974b). Das Treffpunkt-Prinzip als Kommunikationsstrategic und Isolationsmechanismus bei Bohrfliegen (Diptera: Trypetidae). *Entomologia Germanica* **1**, 11–20.

Zwölfer, H. and Bush, G. L. (1984). Sympatrische und parapatrische Artbildung. *Zeitschrift für zoologische Systematik und Evolutionsforschung* **22**, 211–233.

Taxonomic index

Subject index

Printed in the United States
by Baker & Taylor Publisher Services